Bare Bones Geology

For the
Geologically Challenged

ALAN M. CVANCARA

Printed in Victoria, Canada

National Library of Canada Cataloguing in Publication Data

Cvancara, Alan M.
     Bare bones geology / Alan Cvancara.
Includes bibliographical references and index.
ISBN 1-4120-1216-3
     1. Geology—Popular works. I. Title.
QE31.V35 2004   551   C2003-904776-8

TRAFFORD

This book was published *on-demand* in cooperation with Trafford Publishing.
On-demand publishing is a unique process and service of making a book available for retail sale to the public taking advantage of on-demand manufacturing and Internet marketing.
On-demand publishing includes promotions, retail sales, manufacturing, order fulfilment, accounting and collecting royalties on behalf of the author.

Suite 6E, 2333 Government St., Victoria, B.C. V8T 4P4, CANADA

Phone	250-383-6864	Toll-free	1-888-232-4444 (Canada & US)
Fax	250-383-6804	E-mail	sales@trafford.com
Web site	www.trafford.com	TRAFFORD PUBLISHING IS A DIVISION OF TRAFFORD HOLDINGS LTD.	
Trafford Catalogue #03-1594	www.trafford.com/robots/03-1594.html		

10   9   8   7   6   5   4   3   2

To Ella, as several times before:
staunch supporter, writing critic,
travel companion, and
one who tolerated a lot of geology on field trips
masked as vacations.

Among others, photographs in Figures
8-1, 8-3, 11-2, and 14-1 are by the U.S.
Geological Survey.

Contents

Chapter 1
Introduction

Geology deals with how Earth works now and how it has worked in the past. Geologists delve into such diverse features as volcanoes, glaciers, mountains, streams, beaches, and the history of life.

Why Geology for You?

As a resident of Earth you are committed to this planet, and can benefit from some knowledge of it. A grasp of some geology gives you a greater appreciation and understanding of your surroundings, and the capability to perceive nature to a fuller degree. You don't need to know the geology behind beautiful scenery but such comprehension enriches the experience.

Imagine a westward drive, for the first time, toward the Teton Range of Grand Teton National Park just south of Yellowstone National Park in northwestern Wyoming. Your first glimpse of the majestic peaks is unforgettable. As the distance narrows, details begin to show through the first bluish monochrome of the peaks. Beyond the continued marvel of the experience you may wonder: 1) Why do the mountains have no usual foothills? 2) Why are the peaks so jagged? 3) Why do the Tetons seem to run almost straight north-south? With help from this book, you will be able to "read" this landscape and others that pique your interest.

Be aware that geologic features form the underpinnings or foundations of all landscapes. Topping these underpinnings, plants provide interesting cover for rocks and harbor animals.

Earth's persistent clock has ticked for millions of years--even billions of years extending toward the time of Earth's origin. This long history gives you insight into what you may expect in the future. Meteorites have impacted Earth. Can you expect others? Extreme extinctions of creatures have happened in the past. Will they be as extreme again? Glaciers now melt and cause sea level to rise. Will glaciers once again increase in size and cause sea level to drop?

An added benefit from soaking up geologic information in this book and elsewhere is to make you a more informed and responsible voter. You will have gained the means to form sound geologic judgments regarding the exploitation of our finite, nonrenewable resources and protection of our natural environment. If politicians' ideas don't mesh with your analyzed judgments, don't vote for them.

How to Use This Book

This book is for those of you who wish to know some geology--but not too much. Should you become turned on to geology, you may wish to read a more in-depth companion book of mine, *A Field Manual for the Amateur Geologist.* In any case, *Bare Bones,* besides informing you at home, is a user-friendly traveling companion. Its intent is to inform pleasurably.

The book is divided into three parts. Part 1, Eyeballing Landscapes, allows you to recognize and understand landscapes that you travel through. Part 2, Practical Geology: Coping with Geologic Hazards, provides you with information to cope with floods, earthquakes, volcanic eruptions, and the like. And Part 3, Geology to Stir the Brain, gives you the means to think geologically.

"Doing geology" is, in large part, a visual activity. The numerous illustrations in this book attest to this. As a first step, you might flip through the pages, glance at the illustrations and read the captions. Illustrations not given credit are my own.

After glancing at the illustrations you might read Chapters 2 through 9 (the landscapes chapters) and skim Chapters 10 through 19 (the hazards chapters). Skim seriously Chapters 21 (Asking "Dumb" Questions), 22 (Geological Puzzles), and 23 (How Roads Reflect Geology). Now read the chapters that you have skimmed. You may wish to dwell on Chapter 20 (Two Unifying Concepts) when you have more time.

For each chapter I have sifted the real grain from the chaff, and introduce only those technical terms that seem absolutely necessary. Such terms are in *italics*, defined on the spot, and also included in a glossary at the back of the book. I provide informal pronunciations if I believe they might be desirable. Examples: geology (gee-AH-luh-gee) and my surname Cvancara (SWUHN-shuh-ruh); the accented syllable is capitalized.

Besides additional readings listed at the ends of some chapters, along with informative web sites here and there and in a separate section at the end of the book, I recommend your reading *Geotimes*, a news magazine of the earth sciences. It provides you with easily readable accounts of discoveries about our Earth, particularly pointed out in a special highlights issue. This magazine is published by the American Geological Institute, 4220 King Street, Alexandria, VA 22302-1502 (www.geotimes.org).

Geological Strategy for the Outdoor Adventurer

Along with this book, I recommend you bring along a few other items while on your adventurous travels. A simple 10-power magnifier or hand lens unveils the small details of minerals, rocks, and fossils. *Minerals* are naturally occurring inorganic substances with characteristic physical and chemical properties. *Rocks* are naturally occurring substances made up of one or more minerals. *Fossils* are any evidences of past life like a shell, bone, or impression in a rock. A geologist's or mason's hammer provides fresh mineral or rock surfaces to examine.

A small bottle, with dropper, of vinegar or lemon juice helps you identify the carbonate mineral *calcite* or the rock *limestone* because of the fizzing that takes place when you apply vinegar or lemon juice. This also works with the similar mineral *dolomite* and the rock *dolostone* but the fizzing is less intense. Both minerals and rocks also easily scratch with a pocketknife and leave a white powder.

Record your observations in a field notebook to help you remember details and compare the geology from place to place. And don't forget your camera.

When you have time, during your travels, to explore geologic sights and insights, pose questions: Have streams shaped this landscape? Have glaciers sculpted the jagged mountain peaks? Could those tilted layers in the white sandstone cliff have been caused by ancient sand dunes? Could that gray limestone traceable for miles have been originally a lime mud in a tropical sea? Does it have any fossil corals to prove this? A particularly good approach is to ask these questions while exploring national parks and monuments because you can easily test your geological prowess at visitor centers. Before long, more of your geological guesses will be correct. Enjoy the good feeling of having your geologic confidence level soar.

PART 1. EYEBALLING LANDSCAPES

Chapter 2
Landscapes Shaped by Streams

Figure 2–1. Stream sculpted landscape. Part of the Colorado Plateau in southeastern Utah. A white "C" in the low–central part of the photograph marks the confluence of the Colorado and Green Rivers in Canyonlands National Park. The top of the photograph is toward the northeast. This satellite view is from 570 miles (917 kilometers) up. Photograph by the National Aeronautics and Space Administration, courtesy of the U.S. Geological Survey.

Streams fashion most landscapes on Earth. Even in arid regions streams are a dominant sculpting force. Figure 2-1 is an aerial glimpse of part of dry southeastern Utah. Wiggly lines of the Colorado and Green Rivers and their tributaries have dissected the landscape. Even stream valleys etch the mountains and plateaus. All features of the landscape can be seen to relate to the dynamic stream system.

Most stream systems follow the design of a tree, with tiny twigs leading to larger twigs, small branches to large branches, and finally the trunk. In the upper reaches of a drainage system, trickles merge into brooks or creeks, which join with larger tributaries to ultimately attach to the trunk stream. Such is a grand design that forges a dissected landscape with all parts intimately related.

Earth constantly recycles its water. As ocean water evaporates, most rains back to the oceans but 25 percent falls on the land. Most of that 25 percent evaporates again as plant leaves exhale some of the water into the air through their leaves. The rest flows back to the sea in normal streams and in meltwater streams from glaciers, or moves into and through the ground as groundwater. Even with only a relatively small amount of total Earth water, streams are still able to sculpt most of the landscape.

In dry places, such as southeastern Utah, you may be fooled that streams have shaped certain landscapes because they may flow only on occasion, with dry channels much of the time. But a lot of stream work can take place during a rare, single cloudburst.

Streams Erode

Streams erode their channels in a number of ways. *Erosion* is the loosening and movement of Earth material, in this case by streams. The swirling force of running water breaks loose cracked rock of a stream bed, and washes away loose *sediment*, such as sand or the coarser gravel, of a stream bank on the outside of a stream curve or loop. Forceful water carrying sand and gravel grinds away a rocky stream bed like a file abrading metal. *Potholes*, rounded hollows in a rocky stream bed at waterfalls and rapids, attest to this grinding action. Look for potholes in places where stream beds go dry at times. Some streams dissolve the rock over which they flow. A stream with somewhat acid water, for example, can dissolve the limy rock limestone.

Streams Lay Down Sediment

During a major flood, a stream may carry all sizes of sediment from boulders to clay. As stream speed lessens, coarser gravel and sand drop to the bottom first followed by finer silt and still finer clay. *Bars* (Figure 2-2) of gravel and sand in stream channels testify to the laying down or depositing of sediment.

Slower moving streams display meandering channels that curve and loop. In places, loops are cut off from the main channel to form *oxbow lakes*. On either side of a meandering channel lies a broad, flat tract or *floodplain*, which may be covered by water during a flood. As flooding waters recede, silt and clay deposit from the muddy water and gradually build up the floodplain. Sand and silt may build up on either side of the channel after floods to produce low ridges called *natural levees*.

Figure 2-2. Conspicuous sand bars and floodplain (left) of a meandering stream. Little Missouri River, South Unit, Theodore Roosevelt National Park, north-northeast of Medora, southwestern North Dakota. Badlands topography is well developed in the distance.

Where a stream enters a lake or the sea its speed is slowed. The channel splits into many *distributary* channels, the opposite of tributary channels which join a main channel. Carried sediment is dumped into the quieter water to form a *delta*, which, viewed from the air, resembles the Greek letter delta (Δ) in the classic sense of the Nile delta of Egypt. Many deltas don't fit this form because of sweeping and shaping by shoreline waves and currents.

Alluvial fans are the on-land, fan-like counterparts of deltas, especially common in dry regions. A stream emerging from a mountain canyon slows its speed upon reaching flat terrain. Carried sediment drops out and the channel splits into distributaries. During the heavier of infrequent rainstorms, *mudflows*, slurries of fast-moving mud and water, also help build up alluvial fans. Death Valley, California and the Basin and Range region of Nevada are good places to see alluvial fans.

Stream Valleys

Valleys are the most common *landforms*, smaller features of Earth's landscapes, and most are cut by streams. Narrow valleys are narrow at their bottoms--streams fill their bottoms--and are V-shaped in across-valley profile as viewed up-valley or down-valley. Waterfalls and rapids, places where the slope of the stream bed steepens, especially characterize narrow valleys. *Downcutting*, erosion of a stream bed, is significant in deepening valleys. Where downcutting is rapid or in resistant rock, narrow, straight-sided *slot canyons* may develop, such as in Zion National Park, Utah. More often, downslope movement of sediment and rock--by landslides and related movements--and sheet erosion combine to widen valleys to a characteristic V-shape. In sheet erosion, water flows overland in a thin sheet and outside of a stream channel. Streams in such valleys act like conveyor belts to remove the sediment and rock derived from the valley walls by downslope movement and sheet erosion. The Black Canyon of the Gunnison River in west-central Colorado is a classic narrow valley along with the Grand Canyons of the Colorado and Yellowstone Rivers.

Broad valleys are wider than high and have flat bottoms. They widen more than they cut downward, mainly by side-to-side erosion as the stream swings to one side of the valley floor and then the other. The caving in of banks on the outside of curves helps this erosive process. Features found with broad valleys are those already mentioned: meandering channels, oxbow lakes, floodplains, and natural levees. Much of the Mississippi and Missouri Rivers, where not dammed, occupy broad valleys.

How Stream Landscapes Change

Imagine what happens in a stream landscape during a heavy rain. Most of the water flows directly into channels of streams which may be small creeks, brooks, or large rivers. Both stream and sheet erosion loosen and remove soil and sediment--as shown by the muddy water--to mold the landscape. (*Soil* is loose Earth material that forms in place as rocks naturally break up and decay, a process called *weathering*. It differs from sediment, also loose material but which is moved, most often by water and wind.) Strong raindrop splattering also aids in loosening soil and sediment.

Streams, along with their tributaries, are parts of a drainage basin which actually sculpts a landscape. Each stream lengthens its headwater reach by extending upslope toward a drainage divide. (I find humorous some of the terminology geologists use even though I'm part of that crowd.   The uppermost part of a stream is its headwaters but the lowest part is its mouth. Mouths should go with heads!)

Geologists have put together a concept of landscape change by looking at different stages in various places. In a humid region, the landscape undergoes change from one of low relief to high relief and back to low relief. At first, streams flow fast through narrow valleys with waterfalls and rapids. Later, with a landscape of high relief and many tributaries, the main streams become broad with meandering channels and the like. Valley slopes retreat away from stream channels. Finally, as the elevation of a region reaches a controlling level--often the sea but also a lake, reservoir, or resistant rock--the landscape becomes relatively flat once again. Valleys become very broad and floodplains become wider than the width of the belts of meandering streams.

In an arid region, landscape changes are similar but more obvious, and the work of wind may be important. Sedimentary rock layers resistant to erosion, especially of sandstone and limestone, may cap less resistant layers of *shale*, lithified mud, to produce high, flat-topped *plateaus*. These plateaus eventually diminish to *mesas*, which, in time, further waste to isolated hills called *buttes*. High-elevation plateaus and mesas may be called *erosional mountains*. A prominent erosional mountain is Grand Mesa east of Grand Junction in west-central Colorado. It rises to an elevation of more than 11,000 feet (3,350 meters) and is capped by resistant lava.

Without interruptions, these landscape changes tend to take place. But what if something causes an increase in downcutting, caused by greater stream flow or a steeper slope of the stream bed? (I'll not get into the possibilities of these possibilities.)   The development of the landscape will be checked. A couple landforms give us clues to such interruptions. A "rejuvenated" stream in a broad valley cuts into its floodplain to leave bench-like remnants of that

Figure 2-3. Meandering stream with at least three levels of stream terraces. White Earth River, southwest of Manitou, northwestern North Dakota.

floodplain called *stream terraces* (Figure 2-3) on either side of the valley. Terraces are well displayed along the Madison River south of Ennis in southwestern Montana and along the Snake River in Grand Teton National Park, Wyoming. Meandering channels may downcut or incise deeply, while maintaining their shape, to form *incised meanders*. You can see outstanding examples of these in San Juan State Park (called "goosenecks" here) and Canyonlands National Park (Figure 2-4), both in southeastern Utah.

Figure 2-4. Incised meanders. The Loop (largest of the two stream loops shown), Canyonlands National Park, southwest of Moab, southeastern Utah. North is at the top of the photograph, which is by S.W. Lohman, U.S. Geological Survey.

Chapter 3
Landscapes Shaped by Waves and Currents

Waves and longshore currents, driven by the energy of the wind, shape landscapes along sea or lake coasts. Both waves and longshore currents are more powerful in the sea than in lakes because of the greater wave energy. But similar landscapes result at the edges of both. Wave energy tears down coasts in places, builds them up in others.

How Waves and Currents Do Their Work

You, as nearly everyone, are probably fascinated by the incessant onrush and crashing of waves against a shore. As you watch them, bear in mind that wave motion transfers wind energy with little movement of the water itself. Liken this to waves in a wheat field: Their motion sweeps over the field but the wheat stalks are firmly rooted in soil.

You can guage wave energy by the wave length--the distance from the high point of one wave to the high point of another--as well as the wave height--the distance from a high point of a wave to a nearby low point. Waves begin to "feel" or erode the bottom at a depth equal to half of the wave length. So if you estimate that the wave length of two oncoming waves is 600 feet (183 meters), the waves begin to touch bottom at a water depth of 300 feet (91 meters). As waves approach a shore, they crowd together, get higher, oversteepen, and break or topple into *surf*. Waves tend to break at a depth of one to one and one-half times the wave height. Since normal, non-storm, sea waves seldom reach 20 feet (6 meters), the surf zone is usually within a depth of 20 to 30 feet (9 meters).

Breaking waves usually don't rush straight in but arrive at some angle to the shore. They wash sediment up the slope of the beach and backwash it at a slightly different place as they recede. Repeated wash and backwash movement of sediment in curved paths parallel to shore is called *beach drifting*. The net result is migration of sand along the beach.

Waves within the deeper water of the surf, which also strike the shore at an angle, pile up water at first. Then the water is forced to flow parallel to the shore as a *longshore current* which moves sediment by *longshore drifting*. As you might expect, beach drift and longshore drift proceed in the same direction on the same beach.

Waves accomplish their erosional work in ways similar to those of streams. Slamming against rocky shores, they loosen and lift rocks already cracked. They compress air in cavities that pries rocks apart and sprays out with force through blowholes. Waves carry sand and gravel as tools to rasp and grind against sea cliffs. And, in a barely noticeable way, waves dissolve limestone and similar rocks.

Coasts of Mainly Erosion

Figure 3-1. Stacks along a surf-swept rocky coast. Cannon Beach from Ecola State Park, northwestern Oregon.

High-energy waves smash against a coast with projecting points of land (headlands) and cut sea cliffs. The cliffs recede inland as undercutting by the waves causes landsliding, much in the fashion of streams that undercut steep valley walls and widen valleys. Property owners try to protect cliff-edge buildings by building seawalls of concrete or stone at the base of sea cliffs. The seawalls may slow cliff erosion but at a cost. Waves deflected from a seawall increase their energy and may wipe out nearby beaches.

Waves hollow out cavities in sea cliffs called *sea caves*. Caves on opposite sides of a projecting headland may erode through to form a *sea arch*. Collapse of a sea arch forms a *stack* (Figure 3-1), a small

Groin

Figure 3-2. An actively moving beach. Cape Hatteras National Seashore at Cape Hatteras Lighthouse (left), eastern North Carolina. The sand is drifting along the beach toward the bottom of the photograph. Groins (one labeled) help retard sand movement. North is toward the top of the photograph, which is by R. Dolan, U.S. Geological Survey.

rocky island. Stacks also develop where headlands with vertical cracks erode.

Flat *wave-cut benches* or platforms form at the base of retreating sea cliffs as waves pound and grind. Low tide exposes these benches in places. They widen until they reach a critical width which absorbs most of the wave energy.

Coasts of Mainly Deposition

Many coasts are along low plains with gentle slopes toward the sea, shaped mainly by the laying down of sediment by beach drift and longshore drift. Well-developed beaches occur along these coasts, usually of sand but also of gravel--up to the size of boulders--where wave energy is high. What's the source of beach sediment? Mostly from streams that enter the sea. If seaward streams are dammed, nearby beaches tend to shrink.

Beaches exist in a state of constant change (Figure 3-2). They migrate both landward and seaward in response to wave energy, and tend to alter with the seasons. Prevalent storms during fall and winter remove sand, whereas quieter waters during spring and summer re-deposit sand. At La Jolla, California, for example, the cooler seasons present a gravel beach and the warmer seasons, fortunately, usher in a sand beach.

Coast resorts worry about losing "their" beaches from beach and longshore drift. To prevent such loss, some construct groins (Figure 3-2), low walls at right angles to shore. Groins do trap sand on their up-drift sides but encourage sand erosion on their down-drift sides.

In similar manner, some coastal cities worry about drifting sand building up at the entrances to harbors. Similar to groins, a pair of jetties at right angles to shore edge the two sides of a harbor. Some sand accumulates on the up-drift jetty but erodes away on the down-current side of the other jetty. Sand may accumulate in the harbor in spite of the jetties and then must be removed by dredges. At times even humans must accede to Nature!

Where sediment drifting along a beach passes into the deeper water of a bay, a *spit*, a ridge of sand or gravel, forms. Cape Cod, Massachusetts and Sandy Hook, New Jersey are good examples of spits, both curved landward. A large lake example is in northwestern Pennsylvania along Lake Erie's coast and occupied by Presque Isle State Park. If such sediment drift extends a spit to block the mouth of a bay, and seal it from the sea or nearly so, a *baymouth bar* (Figure 12-1) develops.

Adding significantly to the landscape of low coasts are *barrier islands* (Figures 3-2 and 3-3), ridges of sand that extend along low-relief coasts and are separated from the land by lagoons. Besides Cape Hatteras National Seashore, North Carolina, another good example is Padre Island National Seashore, Texas, flanked landward by Laguna Madre. Cities built on barrier islands include Galveston, Texas and Miami Beach, Florida. Barrier islands may form as storm waves breach much-elongated spits to isolate them from land or by the shoreward movement of offshore bars that rise above sea level.

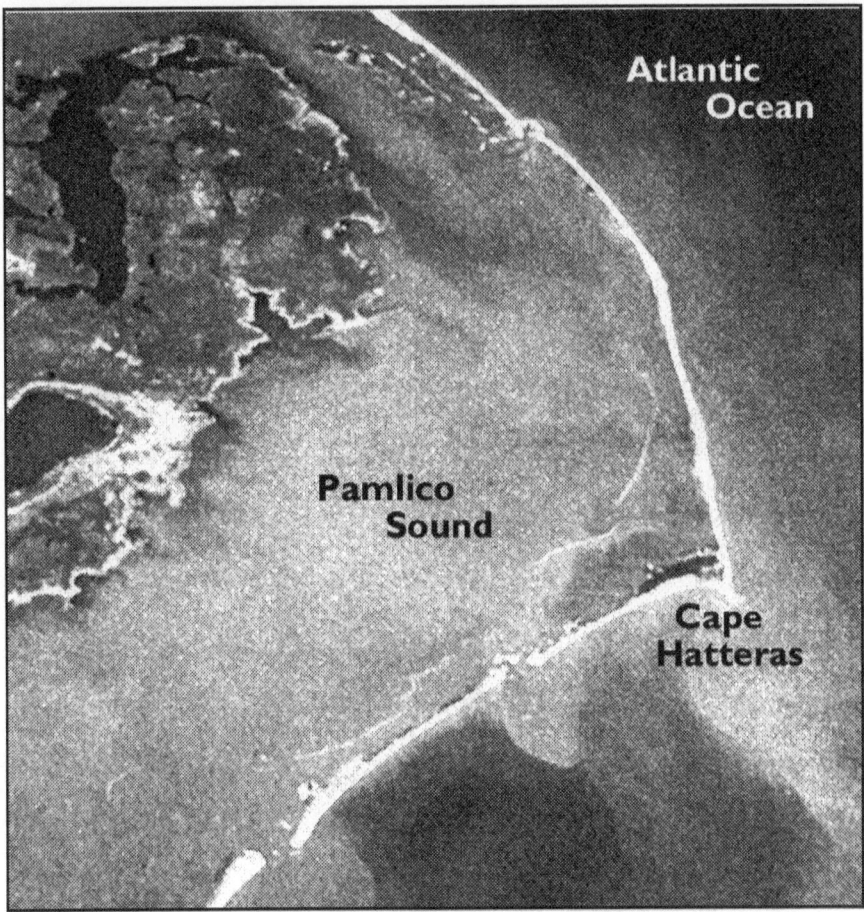

Figure 3-3. Barrier islands. Cape Hatteras National Seashore, eastern North Carolina. Figure 3-2 shows an oblique aerial view of Cape Hatteras. Pamlico Sound is a lagoon shoreward of the barrier islands. North is generally toward the top of the photograph, which is by the National Aeronautics and Space Administration, courtesy of the U.S. Geological Survey.

Drowned and Uplifted Coasts

Many coasts, such as along much of the U.S. Atlantic Seaboard, are considered to be "drowned." Sea level rise (Chapter 19) by melting glaciers has caused flooding of these coasts. *Estuaries*, drowned river mouths, extend considerably landward by the rise in sea level. Chesapeake Bay, Maryland is a notable estuary.

Other coasts show signs of being uplifted by forces within Earth, such as along much of the U.S. Pacific Coast. Sandstone riddled by present-day rock-boring piddock clams tens of feet (or meters) above high tide confirms recent emergence along the northwestern coast of Washington. In other places, uplifted wave-cut benches with stacks now form *marine terraces* that somewhat resemble stream terraces.   Patrick's Point State Park, northern California lies on a terrace with stacks some 100 feet (30 meters) above sea level. In the Palos Verdes Hills south of Los Angeles you can see terraces at several levels, the highest about 1,300 feet (400 meters) above sea level.

Landscape Change Along Coasts

Imagine an irregular coast with projecting headlands and intervening bays, like a drowned coastline originally sculpted by streams. With both shoreline erosion and deposition taking place, and no interruption by uplift or drowning of the coast, the irregular coast tends to gradually straighten. Wave erosion cuts sea cliffs into the headlands and wave-cut benches with stacks begin to take shape. Beaches materialize at the base of the sea cliffs.   More available sediment allows beaches to extend into spits, baymouth bars, and barrier islands. Bays and lagoons fill with sediment. The straightened coastline retreats landward until the wave-cut bench becomes so wide that it curbs wave erosion.

Chapter 4
Landscapes Shaped By Wind

We often think of wind shaping landscapes in deserts but it can mold them in most any place where sediment is loose and dry. Consider wind as a fluid, "thinner" than water, like a thinner oil in your automobile versus a thicker one. Because of its thinness, wind carries only the finer sediment--sand, silt, and clay.

Landscapes Shaped By Wind Deposition

Figure 4-1. Dunes. Northwestern Namib Desert, Namibia, southwestern Africa. The large dune in the background is about 300 feet (90 meters) high. Photograph by E. T. Nichols, U.S. Geological Survey.

Sand dunes, ridges or mounds of sand heaped up by the wind, are scattered within the southwestern U.S. and along both coasts, as well as other places. Good places to see inland dunes are Great Sand Dunes National Park in south-central Colorado, White Sands National Monument in south-central New Mexico, and Killpecker Sand Dunes in southwestern Wyoming. Coastal dunes, formed by onshore winds, develop well at Oregon Dunes National Recreation Area in southwestern Oregon and Sleeping Bear Dunes National Lakeshore on the northwestern shore of the lower peninsula of Michigan. Dunes in places like the Sahara Desert in north Africa are so vast as to create sand seas. Some dunes are, indeed, impressive (Figure 4-1).

Many dunes initiate against an obstacle like a bush or rock that creates a pocket of quieter air on the downwind side. Sand grains roll, slide, and jump on the gentler upwind slope and cascade into "tongues" on the steeper, downwind slope (Figure 4-2). In time, the dunes build up and migrate downwind. On a quiet day, you can always tell the wind direction that aligned a dune by the steeper downwind slope.

Figure 4-2. Front (downwind) face of a sand dune, showing cascading "sand tongues" and adjacent ripples. Oregon Dunes National Recreational Area, south-southwest of Dunes City, southwestern Oregon.

If you could slice across a dune with a huge knife, you would find that its internal layering mostly lines up with the downwind slope. You can apply this information to dunes, millions of years old, that have solidified into rock. At Zion National Park, Utah and Canyon de Chelly National Monument, Arizona you can see lithified dunes cut

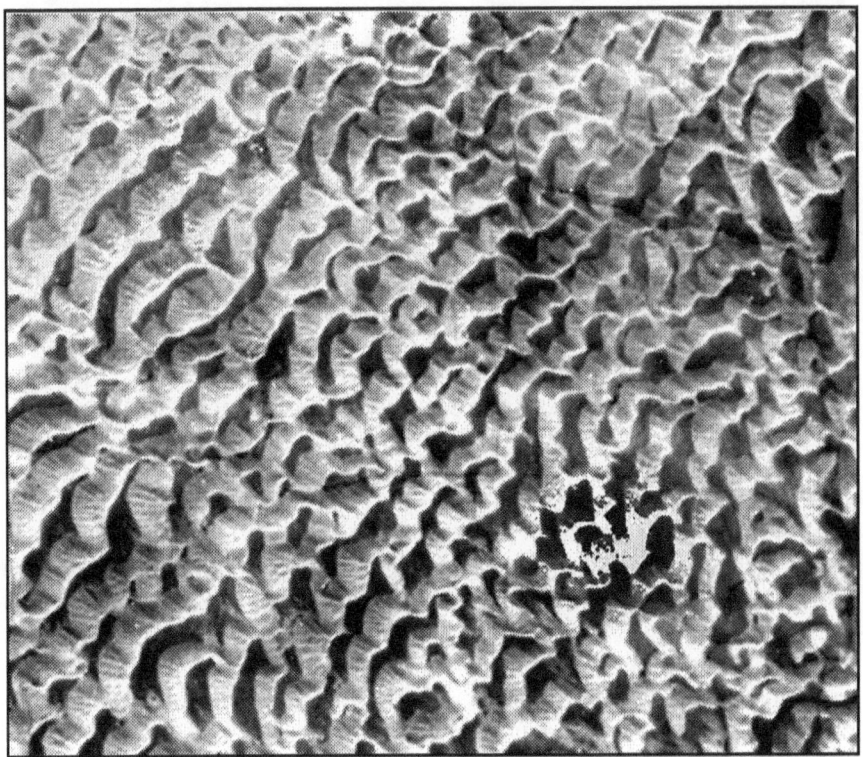

Figure 4-3. Barchan-like sand dunes, aerial view. White Sands National Monument, southwest of Alamogordo, southwestern New Mexico. Most of the crescent-shaped dunes are joined together. The wind has blown from the upper left to the lower right. Photograph by E.D. McKee, U.S. Geological Survey.

through by erosion and reason out wind directions when the sands of these dunes were laid down.

Numerous kinds of dunes exist, depending on wind speed, how much sand is available to shuffle around, and the amount of vegetation that can slow the movement of sand. Some dunes orient at right angles to wind, others parallel the wind, and still others relate to the wind in other ways. Probably most intriguing are *barchans* (bar-KAHNZ) (Figure 4-3), crescent-shaped dunes whose narrowed tips curve around the steeper downwind slope and point downwind. They form in places with moderate winds and little sand and vegetation. On the surfaces of dunes are low *ripples* (Figure 4-2) at right angles to wind direction, like miniature versions of dunes. You can also tell wind direction from wind ripples by their steeper downwind slopes.

It's fun to walk dunes to feel what they are really like. If the wind arises while you're on a dune hike, don't worry unduly. You likely won't get sand in your eyes. Sand grains jump mostly within a height of only about 1 1/2 feet (0.4 meter) and rarely more than 3 feet (0.9 meter).

Sand grains in dunes are well-fashioned by their constant abrasive action with other grains. Under a magnifying lens you can see that most are of about the same size and have their edges and corners rounded off. Most, too, are made up of the glassy mineral *quartz*. Some are of the mineral *feldspar*, white, gray, or pinkish. Others are of dark fragments of various kinds of rock. In some places, black grains streak the sand. If they are of the iron mineral *magnetite*, you can easily pick up these grains with a swipe or two of a small magnet. Most unusual are white, clear, or gray grains of the soft mineral *gypsum* which makes up the dunes at White Sands National Monument.

Dunes interfere with humans at times. In coastal areas, planting beach grass may slow the encroachment of dune sand onto homes. Erecting fences similar to snow fences may also deter this invasion that blocks roads as well as intrudes buildings. Earth-moving equipment with blades may be needed to remove sand that piles up often.

If you live or travel in a snowy region that lacks sand dunes and you wish to experience them, consider snow dunes as substitutes. Yes, most types of sand dunes also take shape in snow. To see a "snow barchan," search in a place with moderate wind and little snow. Be aware that some snow barchans are tiny, only a few to several feet across.

Landscapes Shaped By Wind Erosion

Wind scoops out depressions, *blowouts*, up to several miles across. Many dot the Great Plains. Some blowouts contain lakes or ponds. They often have remnant pedestals of sediment that haven't yet eroded away, an aid to their recognition. A few causes limit the depths of blowouts. One is the *water table*, the top of the zone beneath the surface that is saturated by groundwater (Chapter 6).

Wind can't scoop wet sand. Another is a resistant rock layer, which doesn't yield readily to wind. And, a third, is *desert pavement*, a layer of packed pebbles, left behind with the loose sand removed.

In some deserts abrading sand sculpts narrow ridges from sediment or soft rock. These ridges line up with the prevailing wind direction.

Wind may also sandblast hard rock, armed with sand grain "tools." The results are rock surfaces with grooves, pits, facets, and polish.

Changes in Wind-Formed Landscapes

Wind-formed landscapes change largely because of changes in climate, which bring about changes in wind speed and direction and the amounts of sand and vegetation. If the climate changes from arid to humid, more vegetation grows and dunes become stabilized. Consider the Sand Hills of west-central Nebraska. They're made up of dunes formed thousands of years ago when the climate there was drier. Now they are grown over with vegetation, and many ponds and lakes fill the swales between former dunes. Vegetated dunes can be hard to recognize. Look for jumbled hills with small blowouts lined with sand.

A change from a humid climate to a drier one has the opposite influence. Wind is not obstructed by vegetation and has more sand to play with. Various dunes develop and often shift their positions. Blowouts may become frequent and extensive.

Chapter 5
Landscapes Shaped by
Volcanoes and Lava Flows

Lava, molten rock that issues from vents or cracks at Earth's surface, is the raw material that forms volcanoes and lava flows. (Lava may also refer to the rock that solidifies from molten rock.) Some lava explodes from pent-up gas and may form broken lava fragments such as ash and cinders. *Volcanoes* are conical hills or mountains made up of lava or a combination of lava and broken lava fragments. If higher than about 1,000 feet, they form a second type of mountains, *volcanic mountains.* The first type, erosional mountains, is mentioned in Chapter 2. At the tops of volcanoes are circular or ellipse-shaped depressions--*craters* if less than a mile (1.6 kilometers) across, *calderas* if larger. Craters and calderas form by volcanic explosion or by collapse as lava is withdrawn from below.

Lava flows (Figure 5-1) are outpourings of lava that do not build up into hills or mountains but form low-relief *lava plains*.

Figure 5-1. Lava flow (lower) and cascades at night. Kilauea Volcano, Hawaii Volcanoes National Park, southeastern Hawaii. Photograph by J.B. Eaton, U.S. Geological Survey.

Landscapes Shaped by Volcanoes

Three main types of volcanoes present different profiles to volcanic landscapes. *Shield volcanoes* display a flattened, gently curved profile with slopes of generally fewer than 10 degrees and resemble a warrior's shield. Good examples are Mauna Loa and Mauna Kea in Hawaii. Mauna Loa, Earth's largest shield volcano, juts 30,000 feet (9,100 meters) above the seafloor. But Olympus Mons, a shield volcano on Mars twice as large, easily foreshadows Mauna Loa. Shield volcanoes owe their gentle slopes to free-flowing lava that solidifies mostly into *basalt* (buh-SALT), a black, fine-grained rock rich in iron and magnesium minerals.

Figure 5-2. Cinder cone flanked by lava flows. Lassen Volcanic National Park, west of Susanville, northeastern California. The lighter-colored flow at the left is covered by cinders. Photograph by D.R. Crandell, U.S. Geological Survey.

Cinder cones (Figure 5-2) exhibit much steeper profiles than shield volcanoes, with slopes on the order of 30 to 35 degrees. The steeper slopes result, through the explosive release of gas under pressure, from the build-up of cinders, volcanic ash--finer than cinders, and larger boulder-sized fragments of new and old lava. Cinder cones tend to occur in groups, and lava flows may issue from

Figure 5-3. Composite volcano erupting; Mt. St. Helens, May 18, 1980. Mt. St. Helens National Volcanic Monument, east-northeast of Kelso, southwestern Washington. Volcanic ash, steam, and water rose to 60,000 feet (18,300 meters). Photograph by A. Post, U.S. Geological Survey.

their base.   Examples include Paricutín (pah-ree-koo-TEEN) in western Mexico, Sunset Crater and SP Mountain in north-central Arizona, and Capulin (CAP-yuh-luhn) in northeastern New Mexico. You can walk into the crater of Capulin, nestled amongst four lava flows.

Steeper sloped, smaller, and less common than cinder cones, minor *spatter cones* build up at lava fountains of gas-charged, pasty basalt lava. The belching lava spatters and plasters the sides of the cone as it solidifies.

Most larger volcanoes are *composite volcanoes* (Figure 5-3) with near-summit slopes of about 30 degrees and 5-degree slopes near their bases. A good name, composite, because they're made up of alternating layers of lava flows and broken lava fragments, which reflect alternating quiet and explosive eruptions.   Among the magnificent examples are Mt. Rainier, Mt. St. Helens, Mt. Hood, Mt. Jefferson, Mt. Shasta, and Lassen Peak in the U.S. Cascade Range of Washington, Oregon, and California; Mt. Mayon in the Philippines; Mt. Fujiyama in Japan; and Mt. Etna in northeastern Sicily. Mt. Etna is one of the most frequently erupting composite volcanoes.

Crater Lake, within the Cascades of southwestern Oregon, occupies an old, incomplete composite volcano called Mt. Mazama, and is misnamed. The 1,900-foot- deep lake lies not within a crater but a larger caldera formed by volcanic explosion and collapse. In time, three cinder cones appeared in the caldera, one of which, Wizard Island, rises above lake level.

Composite volcanoes tend to line up within two main belts where you can see impressive volcanic landscapes.   One belt circumscribes the Pacific Ocean, from southern Chile; along the western edges of the Americas to the Aleutian Islands; and southward to Japan, the Philippines, the East Indies, and New Zealand. This belt is aptly named the Ring of Fire. North-south-aligned peaks in the U.S. Cascade Range are part of this belt. Scattered volcanoes also occur within the Pacific region, including those famous in the Hawaiian Islands. The other belt of volcanoes extends from the Mediterranean Sea through the Himalayas and to the East Indies.   Mt. Etna, and another prominent peak, Mt. Vesuvius, are within this belt. These main belts of volcanoes correspond in location to the main belts of earthquakes. In Chapter 20 you will find the theoretical reason for this association.

Do active volcanoes occur elsewhere than on Earth? So far, erupting volcanoes have only been verified by spacecraft on Jupiter's moon Io (EYE-oh).

Landscapes Shaped by Lava Flows

You can recognize flat-lying lava flows by their lobed edges and several surface features. Lava with a wrinkled or ropy surface goes by the Hawaiian name *pahoehoe* (pah-HOE-ee-HOE-ee), and is relatively easy to walk on. Free-flowing basalt lava with considerable gas tends to form this kind of surface. A sharp, jagged flow surface of blocks and cinders is called *aa* (AH-ah), and results from a slow-moving basalt flow with little gas.

Expect other surface features as well. Pressure ridges tend to orient at right angles to the direction toward which a flow has moved, and often display cracks at their crests. They form as the partly rigid flow rock is squeezed while the interior of the flow remains fluid. Squeezeups resemble pressure ridges but are moundlike. Lava that moves at moderate speed may be confined to lava channels which are open at the top or lava tubes if roofed over. Drained, cooled tubes become long caves that may contain ice the year round. The roofs of tubes collapse to channel-like depressions with roof remnants that form natural bridges or arches.

At Craters of the Moon National Monument in south-central Idaho you can witness all of these features associated with lava flows, along with cinder cones and spatter cones. The latest eruptions took place only about 2,000 years ago. Another good place to see lava flows and related details is at El Malpais (ell-mahl-pie-EES) National Monument in west-central New Mexico. One lava tube system is 17 miles (27 kilometers) long.

Lava plains are not limited to Earth. Low-lying dark areas of the moon, maria (MAR-ee-uh), which translate to "seas" because early thinkers believed them to be seas, are actually basalt lava plains. This notion has been confirmed by rock samples recovered by Apollo missions. Winding rilles--trenches or valleys--may be lava channels or collapsed lava tubes. Maria have also been recognized on Mercury, Venus, and Mars.

Some lava flows, where eroded through and viewed from the side, seem to be made up of tightly packed posts or columns called *columnar structure*. As a flow cools, it contracts and sets up a system of many-sided cracks similar to mud cracks in a drying pond. Mud cracks, though, form thin "tablets" rather than extended columns. If shrinkage cracks are symmetrically arranged, six-sided columns are the outcome. Columnar structure prevails in many-stacked basalt lava flows in the Columbia-Snake River Plateau of Washington, Oregon, and Idaho. Devil's Postpile National Monument in eastern California also features columnar structure in basalt (Figure 5-4).

Figure 5-4. Columnar structure in basalt. Devils Postpile National Monument, west of Mammoth Lakes, eastern California. Photograph by the U.S. Geological Survey.

How Volcanic Landscapes Change

Volcanic landscapes change as volcanoes erode. Cinder cones erode faster than other volcanoes because of their makeup of broken lava fragments. In time, the volcano exists only as a sharp peak or spire of more resistant rock that filled the volcano's throat, and is called a *volcanic neck*. A classic example is Ship Rock (Figures 22-6 and 22-7), almost in the northwestern corner of New Mexico. Others include Agathla Peak in northeastern Arizona and Boar's Tusk in southwestern Wyoming. Most geologists also consider Devils Tower in northeastern Wyoming as a volcanic neck, which displays columnar structure well.

Often radiating from volcanic necks are *dikes*, flat-sided bodies of rock that cut across other rocks. If vertical and more resistant than surrounding rocks, dikes create wall-like ridges (Figure 22-6) that enhance the character of volcanic landscapes. These kinds of dikes formed as molten lava forced its way through cracks toward the surface. If formed at considerable depth, dikes are made up of rocks other than lava.

Landscape change of lava fields involves what geologists call inverted topography; that is, low areas gradually evolve into areas of higher elevation. First, imagine a condition whereby relatively little lava extrudes and stays within stream valleys. The lava dams valleys to form lakes upstream. Diverted water cuts new channels along the edges of the flows, more resistant to erosion than the underlying softer rocks. In time, the once low-lying lava fields evolve to irregular, higher-elevation lava-capped plateaus and mesas, which may further waste away to buttes. Any volcanoes in the terrain reduce to spired volcanic necks, perhaps associated with sharp dike ridges. If much lava extrudes on a stream landscape, lava flows fill the valleys and cover the drainage divides between the valleys as well. Some streams cut around the edges of the lava plains, others, with greater slopes, cut through the lava. As before, the lava plains become an elevated lava plateau as softer underlying rocks are worn down and away. The plateau may be further dissected into mesas and buttes. Mentioned above, one such plateau is the Columbia-Snake River Plateau in Washington, Oregon, and Idaho.

Chapter 6
Landscapes Shaped by Groundwater

Groundwater is water beneath the ground surface. It fills pore space between grains in sediment and sedimentary rock and within the cracks of all kinds of rock.

Groundwater is about 60 times more plentiful than water in streams and lakes. About 15 percent of the rain and snow that falls sinks into the ground to form all groundwater.

Groundwater erodes rocks by dissolving them: mainly limestone (and the related rock dolostone) but also rock gypsum and rock salt. In time, groundwater re-deposits these same materials somewhere else.

The Water Table

Groundwater doesn't saturate the shallow underground, except maybe directly after a rain. Just below the ground surface is a zone of mostly air with some water. Below that is a zone saturated with water, the top of which is the water table. Most streams, lakes, springs, and marshes are at the level where the water table intersects the ground surface. The position of the water table tends to mimic the lay of the land--higher under the hills and lower in the valleys, and rises and falls with wet and dry periods.

Groundwater moves from places with a high water table and high water pressure to areas with a lower water table and lower water pressure. Movement is along curved paths, downward as well as sideways. Groundwater flow is much slower than stream flow, on average a few feet to many feet (few to many meters) a day. For flow to occur, rock openings, be they pores or cracks, must be well connected. An *aquifer* (from the Latin *aqua*, water and *-fer*, bearing) is a rock or sediment layer that bears groundwater and through which the water moves with ease.

Some people have the misleading idea that groundwater usually moves in underground streams. Only in rare instances, does underground water flow in caverns or tunnels. Here, though, flow speed may approach that of streams at the surface.

Landscapes of Groundwater Erosion

Landscapes shaped by groundwater erosion occur mainly in limestone because the more easily dissolved rock gypsum and rock salt are rare at Earth's surface. Acid groundwater moves along cracks and rock layering surfaces to dissolve limestone, mostly below the

water table. Water combines with carbon dioxide--which gives the
fizz to carbonated beverages--to form carbonic acid that does the
dissolving work. Caves and the larger caverns gradually form.

The dissolving of limestone produces a peculiar kind of terrain
called *karst topography*, named after the Karst region in western
Yugoslavia. Streams in a karst region are not part of a drainage
network, and those present tend to disappear underground where
they flow into underground chambers. *Sinkholes* (Figure 6-1),
solution cavities open to the sky, pepper the surface. They form by
collapse of caves or caverns or by the dissolving of limestone at the
surface. The throats of sinkholes may plug with mud to form ponds
and lakes. Sinkholes merge and enlarge to form long, blind-ended,
solution valleys.

Figure 6-1. Sinkhole with collapsed home. Bartow, central Florida. The
sink was 520 feet (158 meters) long, 125 feet (38 meters) wide, and 60 feet
(18 meters) deep. Photograph by the U.S. Geological Survey.

You find karst topography mostly in humid and temperate
places. The best-known in the U.S. is in Kentucky, southern Indiana,
and central Florida. Mammoth Cave National Park rests within the
karst region of southwestern Kentucky. Other places include Jamaica
and Puerto Rico. In tropical places, as in southeastern China, the
karst is often reduced to conical or tower-like hills (Figure 6-2) that
resemble French loaves of bread on end.

Landscapes of Groundwater Deposition

Figure 6-2. Tower-like hills as part of karst topography. Li River in foreground, south of Guilin, southeastern China. Photograph by the U.S. Geological Survey.

Surface landscapes sculpted by groundwater deposition are subtle. In places, such as Yellowstone National Park with hot springs and *geysers* (Figure 6-3), hot springs that erupt with steam, heated groundwater issues to the surface in shallow basins. Groundwater deposits rock material into cones, mounds, and terraces as carbon dioxide expels, the water evaporates and cools with a drop in pressure, and algae aid in the process. The water lays down limy or silica-rich rock depending on the rock that groundwater dissolved at depth.

In places, such as Petrified Forest National Park in northeastern Arizona, groundwater deposition has played a part in the formation and preservation of petrified wood (Figure 6-4), strewn about badlands slopes. When the original wood logs were buried, perhaps by hundreds of feet of sediment or rock, groundwater filled in the space between wood cells or replaced them with silica to form various kinds of quartz. This hardened the logs and made them resistant to erosion.

In dry, shallow basins, evaporating groundwater deposits *alkali*, white salts often rich in sodium that you can see from afar. Alkali soils are generally poor for plant growth.

Figure 6-3. Geyser.   Riverside Geyser, Yellowstone National Park, northwestern Wyoming. The hot, erupting geyser exists at the edge of a cool stream. Photograph by W.B. Hamilton, U.S. Geological Survey.

Water-filled caves and caverns drain as the water table lowers. This may take place if nearby stream valleys are cut deeper and water-table lowering follows in response to the valley deepening, or if a region is uplifted. Now filled with air, caves and caverns become places of deposition and form underground landscapes. Water drips, usually through a crack in the ceiling, and lays down limy *dripstone* (Figure 6-5). Icicle-like *stalactites* (stuh-LACK-tights) of dripstone

Figure 6-4. Petrified logs. Petrified Forest National Park, east of Holbrook, northeastern Arizona. Badlands terrain is evident in the distance.

hang from the ceiling. The letter *c* in the word "stalactite" can remind you of the word ceiling. Thicker *stalagmites* (stuh-LAG-mights) stick up from the ground or cavern floors, and can be remembered by the *g* in the word for ground. Some stalactites and stalagmites join during the dripstone-forming process to create columns. In some caves, water flows--instead of drips--in a thin film on cave floors or walls to form sheet-like or ribbon-like *flowstone*. Among the many underground landscapes are Mammoth Cave, Kentucky, already mentioned, and Carlsbad Caverns, New Mexico. Refresh yourself in such places on a hot, summer day: Sense the cool, dark, wet landscape and marvel at the subterranean scenery.

How Groundwater-Shaped Landscapes Change
Imagine a limestone terrain dissected by streams. When the stream valleys become deep, groundwater moves toward them along rock fractures and rock layering surfaces and dissolves the rock along the way. In time, caves become numerous, then enlarge to caverns to infest the underground, and disappearing streams form on the surface. Collapse of cave and cavern roofs produce sinkholes, which proliferate; many plug to form ponds and lakes. Many of the sinkholes enlarge and merge into solution valleys. When most of the

Figure 6-5. Dripstone. Carlsbad Caverns National Park, southwest of Carlsbad, southeastern New Mexico. A column has formed on the left. On the right, a stalactite (from the ceiling) has almost joined a stalagmite (up from the ground) to form another column.

limestone is dissolved away, most traces of an original stream landscape are gone, and only scattered hills remain. Only when all of the limestone has been removed, to expose a less easily dissolved rock, does a normal stream landscape once again appear.

Chapter 7
Landscapes Shaped by Glaciers

Figure 7-1. Valley glacier. Sherman Glacier, Chugach Mountains, southern Alaska. Dark stripes on the glacier are moraines. Photograph by the U.S. Geological Survey.

Glaciers are masses of ice that move--actually flow--under the pressure of their own weight. How can that brittle ice in your glass flow? Glacier ice below a depth of 100 to 200 feet (30 to 61 meters) behaves like a mobile, yielding substance that has the capability to flow. Glaciers move about as fast as groundwater, on the order of less than an inch (2.5 centimeters) to several feet (meters) in a day.

Figure 7-2. U-shaped glacier valley. Astoria River Valley, Jasper National Park, west-central Alberta, Canada.

Present-Day Glaciers

Glaciers today occupy only about 10 percent of Earth's land surface. But within the last two million years they covered up to 30 percent of the land's surface. In North America, glaciers spread over essentially all of Canada and Alaska and as far south as the Ohio River.

Glaciers are of two types, *valley* and *continental*. Valley glaciers are confined to mountain valleys (Figure 7-1). Most glaciers in North America are valley glaciers. Continental glaciers--called ice sheets if larger, ice caps if smaller--are not confined to valleys and cover large areas of the land surface. Ice sheets occur only in Greenland and Antarctica, which contains 90 percent of all glacier ice. On these continents the ice is up to10,000 feet (3,050 meters) or more thick--about *2 miles* (3.2 kilometers) or more! The Barnes Ice Cap (Figure 7-3) is a classic ice cap on north-central Baffin Island in

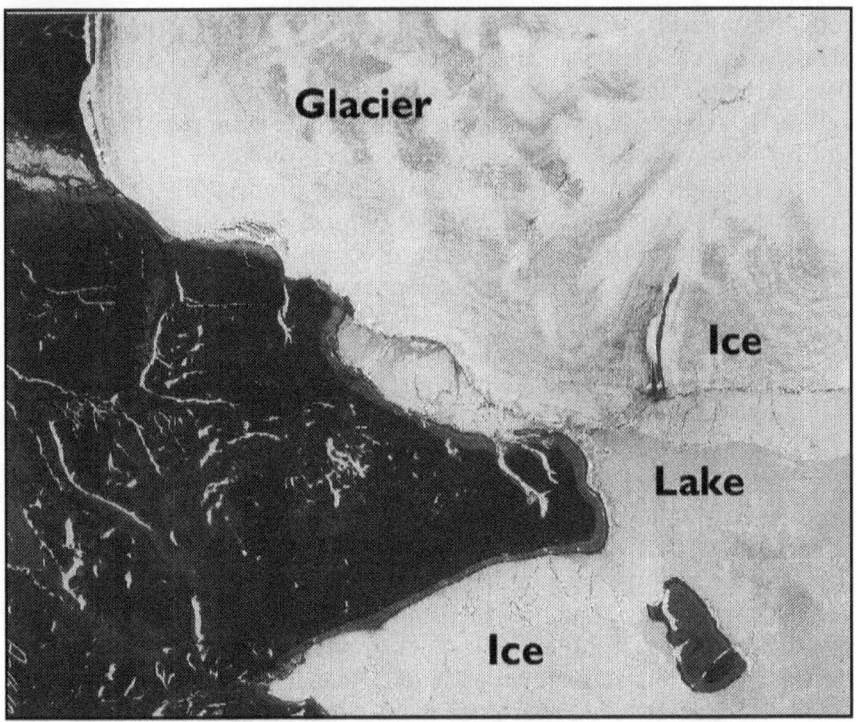

Figure 7-3. Margin of a continental glacier. South edge of the Barnes Ice Cap, central Baffin Island, Canada. Ice of Blanchfield Lake abuts the glacier in the lower part of the photograph and surrounds an island. Dark areas are rocky terrain with little or no snow cover. Aerial photograph A16103-69, Her Majesty the Queen in Right of Canada, reproduced from the collection of the National Air Photo Library with permission of Natural Resources of Canada.

northeastern Canada. This ice cap is one of the last remnants of once-extensive continental glaciers in North America.

Glaciers form where more snow accumulates than melts. Pressure from the weight of accumulating snow transforms the snow into glacier ice. Today, most glaciers are melting rather than building up.

Landscapes of Valley Glaciers

Like water and wind, glacier ice both erodes and deposits rock material, and glacier-produced landscapes evolve from both kinds of geologic work. Rock particles that glaciers carry grind against the rock through which they pass. The abraded rock surfaces become grooved, scratched, and polished. Glaciers also erode as they pluck rock fragments away.   Meltwater in cracks freezes and expands. Blocks of rock are worked loose, frozen to the glacier, and plucked free as the glacier moves down the valley. Similar to streams eroding valleys, glaciers get help molding their valleys from rock falling and sliding from valley walls.

Erosion   by   glaciers   creates   a   fascinating   landscape. Conspicuous are U-shaped valleys (Figure 7-2), in across-valley profile, and often jagged mountain peaks and ridges and steep cliffs (Figures 7-1 and 8-1). The angular peaks and ridges of Grand Teton National Park in Wyoming and Glacier National Park in Montana, as well as the scenery of Yosemite National Park in California, are the outcome of erosion by valley glaciers. Melted tributary glaciers leave *hanging valleys* above the main U-shaped valley. Waterfalls, like Bridalveil Fall in Yosemite, may plunge from these hanging valleys. At the upper ends of valleys where glaciers originate are half-bowl-like depressions called *cirques* (SIRKS). Lakes may occupy cirques and other rock basins within valleys hollowed out by glaciers.

The laying down of rock debris also contributes to valley glacier landscapes. Much of the debris piles up into *moraines*, most often ridges of a jumbled mix of clay, silt, sand, and gravel up to boulder size. Moraines form at several places. Some develop at the sides of a glacier. Side moraines from a main glacier and a tributary glacier join to form a moraine in the middle of the main glacier which gives a striped effect from the air (Figure 7-1). At the snouts of glaciers arc-like moraines mark the places of greatest ice advance. "Snout moraines" may dam up meltwater to form lakes up-valley from them.   On the valley floor another kind of moraine presents an undulating terrain of low hills and enclosed depressions. Beyond the glacier's snout, meltwater and rainwater "wash out" sand and gravel into a nearly flat *outwash plain*. Streams that flow on the glacier's surface or in tunnels within or at its base lay down sand and gravel in twisting ice channels. As the ice melts, channel sediment is let down in snaky ridges, *eskers*, upon the hilly landscape.

Landscapes of Continental Glaciers

Landscapes shaped by continental glaciers are unspectacular and often difficult to recognize. But the clues are evident if you look with care. Stream drainage systems, especially where glacial sediment is thick, tend to be poorly developed or absent. The terrain is hilly or undulating with enclosed depressions that may be occupied by ponds, lakes, marshes, or bogs. Expect to see many boulders on the surface, some as large as houses. In non-mountainous terrain only ice can move such mammoth blocks of rock. Take a closer look to find that the boulders are foreigners--their makeup indicates a distant source: limestone or granite boulders, for example, in a terrain where the closest rocks are of sandstone or shale.

Erosion by continental glaciers produces similar features as for valley glaciers--grooved and scratched rock surfaces. Where ice is thick, mountain peaks and ridges may be rounded off. Much of northeastern Canada displays landscapes with these features. Stream-lined rock knobs, abraded by passing ice with rock-debris tools, have steeper slopes on their down-glacier sides, and reveal the direction of ice movement. (Down-glacier signifies the direction toward which a glacier moved.)

Large rock basins may be scooped out by distinct ice lobes along the edges of ice sheets. The Great Lakes occupy basins that were along former stream valleys and widened by ice lobes. Scooping by ice lobes also created the basins for the Finger Lakes of west-central New York.

Many features of deposition compose the landscapes of continental glaciers. Moraines are similar to those of valley glaciers except for the lack of side moraines and those in the middle of valley glaciers. Outwash plains and eskers are also similar. Streamlined hills, usually of glacier-deposited sediment, resemble streamlined rock knobs. These *drumlins*, however, have *gentler* slopes in the direction of ice movement. Drumlins resemble inverted teaspoons (minus the handles) from the air. They are well-shown east of Great Slave Lake in northwestern Canada, southeastern Wisconsin, and west-central New York north of the Finger Lakes.

Some of the steeper-sided depressions, *kettles,* in undulating moraines or outwash plains, have a peculiar origin. Ice blocks that separated from an ice sheet are buried and insulated for a time. When they melt, kettles form and may fill to become ponds, lakes, and bogs.

How Glacier-Shaped Landscapes Change

In mountainous regions valley glaciers tend to occupy former stream valleys which they deepen, straighten, and widen. Side and middle moraines stripe the flowing ice. Upon melting of the ice, U-shaped valleys can be seen to replace former V-shaped stream valleys. Tributary glacier valleys, unlike tributary stream valleys, "hang" above the main glacier valleys. Jagged peaks and ridges often look down on the broad valleys below. Once suspended side and middle moraines are let down to join other moraines on the valley floors. In time, streams may occupy the glacier valleys.

Continental glaciers tend to grind and scoop away stream-sculpted terrain because such terrain is prevalent on Earth's land surface. In places, when the ice melts, scoured terrain remains. In other places, various glacier-related features of deposition are left behind: moraines, outwash plains, eskers, drumlins, and foreign boulders. Kettles dot the landscape. Where glacier sediment is relatively thin, streams may, in time, produce a drainage network to renew molding the once-glaciated landscape.

Continental glacier landscapes also change with diversion or damming of meltwater and rainwater by the ice. Before glaciers appeared on the scene, most streams in central North America flowed northeast to the Arctic and Atlantic Oceans. The upper Missouri and Ohio Rivers followed this pattern. Glaciers that spread over the northern part of the continent diverted much of the stream flow to the Gulf of Mexico. The Missouri and Ohio Rivers were diverted southward as well, and their modern courses are at about the edges marking the greatest extent of former ice sheets.

Continental glaciers also dammed northerly-flowing streams in North America to create impressive ice-edged lakes. The largest was Lake Agassiz (AGG-uh-see), formed by blockage of the Red River of the North. This former lake is now marked by an extensive fertile lake plain that covers eastern North Dakota, western Minnesota and parts of Ontario, Manitoba, and Saskatchewan. Lakes Winnipeg, Winnipegosis, and Manitoba remain as testimony of this once great lake. In this region the Red River still flows north and tributaries join it to cut into the lake plain.

Chapter 8
Landscapes Shaped by Landslides

Figure 8-1. Rockslide mass covers a valley glacier. Sherman Glacier, Chugach Mountains, southern Alaska. The rockslide, set off by the March 27, 1964 earthquake, had its source at the fresh scar on Shattered Peak in the central distance.

Landslides are the rapid downslope movements of rock and rock debris. If you can see the movement, it's rapid. Rock debris includes any loose material--soil, sediment, and rock fragments. Landslides prevail in hilly or mountainous regions. Gravity's pull is the driving force. As mentioned in previous chapters, landsliding contributes to erosion by streams, waves, and glaciers. Even in the movement of sand dunes, a kind of landsliding plays a part: Sand cascades down the downwind faces of dunes as they march overland.

Figure 8-2. Rockslide on coastal cliff. Loma Prieta, north of Ft. Funston, central-western California. Photograph by S.D. Ellen, U.S. Geological Survey.

Kinds of Landslides and Their Landscapes

Geologists don't always agree on the kinds of landslides. I'll keep the grouping simple.   Let's consider that all landslide movement--involving both rock and rock debris--either falls, slips or slides, or flows.

A *rockfall* involves rock that breaks off and falls directly or bounces off a cliff on its way down. You can see rockfalls along stream-cut canyons, glacier valleys, or wave-cut cliffs. Falling rock debris produces the debris counterpart of a rockfall, a *debris fall*. Both rockfalls and debris falls produce *talus* at the base of cliffs, half-cone-like accumulations with steep slopes.

In a *rockslide* (Figures 8-1 and 8-2) a mass of rock slips or slides parallel to a slope along surfaces of weakness. Rockslides are the most devastating of landslides. A major earthquake triggered the 1959 Madison Canyon rockslide, just west of Yellowstone National Park. A rock mass, which slid from the south canyon wall and reached more than 400 feet (122 meters) high on the opposite north wall, dammed up the Madison River to create Quake Lake. The slide buried at least 26 campers. Weathered metamorphic rock, with stress-induced layering inclined toward the canyon, was ready for dislodgement. Overlying cracked, dolostone, a rock similar to limestone, was also tilted toward the canyon.

The two other largest rockslides in North America that happened within historic time are the Gros Ventre (grow-VAHNT) (1925) and Turtle Mountain (1903). The Gros Ventre rockslide, just southeast of Grand Teton National Park, has a similar setting to that of the Madison Canyon rockslide: It slid from a south canyon wall and dammed up a river, the Gros Ventre. This river oversteepened the lower valley slope. Heavy rain and melting snow saturated sandstone and shale, both inclined toward the valley. The water-weighted sandstone slid down on the greasy, water-lubricated shale. The Turtle Mountain rockslide near Frank in southern Alberta killed 66 people. Movement along steeply inclined cracks in limestone created this catastrophe.

Counterparts to rockslides are *debris slides*. A snow avalanche is similar to a debris slide. Sometimes rock debris and rock move as distinct masses that rotate backward and maintain their general size and shape; these slides are called slumps. Cliffs behind the slumps testify to their sources.

A telltale landscape reveals rockslides and debris slides. First you notice a slide scar or cliff at the source of the landslide if the source is not concealed by plant cover. Below the scar or cliff is rather chaotic terrain. Arc-like ridges or hills alternate with elongate or near-circular depressions creating a landscape similar to that produced by certain glacial moraines. In the case of slumps, ponds or lakes may fill depressions at the base of slide-formed cliffs.

Debris flows, made up of a mixture of rock debris, mud, and water, flow instead of slide. They have the consistency of a thick fluid, something like freshly poured cement or maybe like a thick milkshake. Mudflows are debris flows that are made up of much mud

Figure 8-3. Mudflow destruction in logging camp. North Fork Toutle River Valley, about 17 miles (27 kilometers) down-flow, northwest of Mt. St. Helens, east-northeast of Kelso, southwestern Washington. This mudflow resulted from the eruptions of Mt. St. Helens in 1980.

and are more fluid and move faster than regular debris flows. I mentioned them in Chapter 2 in relation to alluvial fans. Mudflows are capable of carrying huge boulders as well as houses and vehicles.

Volcanic eruptions often set off especially destructive mudlflows. When Mount St. Helens in southwestern Washington erupted in 1980, mudflows--as slurries of volcanic ash and water--rushed down stream valleys. Water came not only from rainfall and melting snow but from melted glacier ice. Most destructive was the mudflow that hurried down the North Fork of the Toutle River off the volcano's northwestern slope (Figure 8-3).

Debris flows produce landscapes a little different from those of landslides. At the lower edges of debris flows are hilly or ridged lobes--something like the lobes at the edges of glaciers--that signify flowage. Cuts in hillsides reveal where flows originate. Some slumps change to debris flows in their lower parts.

Some kinds of downslope movement are too slow to witness, and are not, in a strict sense, landslides. One type, well-named, is *creep,* the very slow downslope movement of rock debris. So slow that average movement is less than an inch (2.5 centimeters) per year. Creep predominates in temperate and tropical climates on slopes with good plant cover. How can you tell creep is in progress? The clues

Figure 8-4. Aspen tree trunks bent downslope by creep.

are numerous: the bending downslope of the lower parts of tree trunks (Figure 8-4); the downslope tilting of power line poles, fenceposts, gravestones, and retaining walls; the downslope slinging of fences and roads; and the downslope bending of the upper edges of steeply inclined rock layers.

Alternate freezing and thawing enhances creep. Expansion of ice upon freezing forces rock debris particles upward at right angles to a slope. With thawing, these particles are let straight down. Repeated freezing and thawing causes tiny debris particles to migrate downslope in small up-and-down increments that add up to noticeable downslope movement over time.

Chapter 9
Landscapes Shaped by Deformed Rocks

Figure 9-1. Bent and flat-lying sedimentary rock layers. North of Mexican Hat, southeastern Utah. The San Juan River cuts through the strata. **Puzzle 1, Chapter 22.**

Earth has a long history of rock deformation. Rocks have been tilted, bent, and broken. Rocks must be buried at considerable depth to become pliable and yield to tilting and bending by Earth forces. At lesser depth, rocks are brittle and yield to stresses by breaking. Compare these kinds of rock deformation with the deformation of glacier ice, mentioned in Chapter 7: the ice being brittle near its surface but having the ability to flow at a thickness of 100 to 200 feet (30 to 61 meters). In Chapter 20 I'll touch on what causes the Earth

Figure 9-2. Tilted rock layers overlain by flat-lying layers. Grand Canyon National Park, northwestern Arizona. View is to the north from Moran Point. The Colorado River is at the bottom. Photograph by E.D. McKee, U.S. Geological Survey. **Puzzle 2, Chapter 22.**

forces necessary to deform rocks.

Landscapes of Tilted and Bent Rock Layers

All tilted or bent rock layers (Figures 9-1 and 9-2) were once horizontal or nearly so. This is because rock layers were once sediments laid down by streams, glaciers, and the wind or deposited in the sea. Some rock layers, of course, are formed from lava flows. So the horizontal reference allows us to judge the amount of tilting or

Figure 9-3. Hogback. Red Canyon, south of Lander, west-central Wyoming. A stream twists on the back slope.

bending of rock layers.

Landscapes derived from deformed rock layers are related to uneven erosion of various types of rock. Some rocks erode faster than others. These differences in erosion are mentioned in Chapter 2 whereby resistant cap rocks, such as sandstone, lead to the formation of plateaus, mesas, and buttes. Be aware that in arid climates, sandstone, *conglomerate*--lithified gravel, and limestone are all quite resistant to erosion whereas shale is not. In humid climates both limestone and shale tend to erode readily.

How do tilted rock layers characterize the landscape? Resistant rock layers tilted at a considerable angle form *hogbacks* (Figure 9-3), ridges with about equal slopes on both sides. *Cuestas* (KWESS-tuhz) are similarly-formed ridges with a steep slope on one side and a gentle slope on the other. Remember that hogbacks and cuestas differ from other ridges in that tilted resistant rock layers are involved.

Figure 9-4. Cane Creek anticline. Canyonlands National Park, southwest of Moab, southeastern Utah. View is from Dead Horse Point. Layers of the anticline are slightly bowed upward. The La Sal Mountains loom in the distance. Photograph by S.W. Lohman, U.S. Geological Survey.

Rock layers bend into two main kinds of folds: *anticlines* (ANT-uh-klynz) (Figure 9-4) and *synclines* (SIN-klynz). Anticlines are up-buckled folds or folds that arch upward. To perceive anticlines, place a stack of sheets of paper on a table. Push opposite edges of the sheets toward each other with both hands so the sheets buckle upward in the center. You now have an anticline; each sheet can simulate a rock layer. Synclines are down-buckled folds or folds that bend downward. If you take the sheaf of paper off the table and buckle the sheets downward, you have a syncline.

Anticlines and synclines usually occur together. As they erode, resistant rock layers form long ridges and nonresistant rock layers form long valleys. Hogbacks and cuestas are most often parts of the sides of eroded anticlines and synclines. If the folds are also inclined downward at their ends, upon erosion they produce a zigzag pattern of ridges and valleys, best seen from the air. The Ridge and Valley region of the eastern United States (Chapter 23, Southwest-central Pennsylvania), which extends from the St. Lawrence Valley to central Alabama, is made up of fold-produced ridges and valleys that are also broken in places. Such ridges are high enough to classify as *folded mountains*. The mountains are not folded but are made up of folded, and also usually broken, rock layers.

Roughly circular, or at least not decidedly elongate, eroded anticlines, are called *domes*. Imagine the strata like a set of upside

Figure 9-5. Stubby hoodoos in sandstone. Colorado National Monument, west of Grand Junction, west-central Colorado.

down nested bowls. Concentric hogbacks and cuestas have their steeper slopes facing inward toward the center of domes. The Black Hills, like a gigantic Earth blister, is a well-known dome (Figure 23-5), again best seen from the air. With similar eroded shapes like those of domes are *basins*, near-circular synclines. With synclines, strata are arranged like nested bowls right side up. Because strata are inclined toward the centers of basins, concentric hogbacks and cuestas have their steeper slopes facing outward from their centers.

Landscapes of Broken Rocks

Rocks are broken by *joints* or by *faults*. Joints are cracks in

Figure 9-6. Natural arch (right) and one in the making. The Windows Section, Arches National Park, north of Moab, southeastern Utah. A hoodoo stands guard on the left.

rocks along which little or no slippage has taken place. Faults, on the other hand, are cracks along which rock blocks have slipped, in some cases, great distances.

Joints, best seen from the air, tend to occur in sets of many and in two or more directions so as to divide rocks into distinct blocks where they intersect. In Colorado National Monument in west-central Colorado, Bryce Canyon National Park in southern Utah, and elsewhere, closely spaced intersecting joints allow erosion to take place from several sides and form rock pillars or columns. Some of these, sculpted into fantastic shapes even resembling human forms, are called *hoodoos* (Figure 9-5).

In Zion and Arches National Parks in southern Utah, vertical joints cut through sandstone whose sand grains are cemented by the mineral calcite. Groundwater and surface water seep along the joints to dissolve the calcite, and widen the joints. Wind and gravity also contribute to the erosion which sculpts the jointed sandstone into rock fins. Further erosion attacks the fins from the sides to form alcoves which are eventually cut through into natural arches (front cover of this book and Figure 9-6). In Zion many alcoves exist but few arches. In Arches, natural arches are legion. If a stream is involved in cutting through a rock fin or similar narrow body of rock, a natural bridge forms (Figure 9-7).

Figure 9-7. Natural bridge. Rainbow Bridge, Rainbow Bridge National Monument, west of Mexican Hat, south-central Utah. View is upstream. Photograph by W.R. Hanson, U.S. Geological Survey.

Movement of rock masses along faults takes place up or down--vertically or at some lesser angle--or horizontally. Some movement is due to Earth forces squeezing toward each other (Figure 9-8), whereas other movement is the result of forces pulling apart from one another. Mountains created by these kinds of faulting are called *fault-block mountains*--mountainous blocks that have slipped along one or more faults. In the Rocky Mountains many of the ranges have formed by squeezing of the rocks into anticlines, which have been later faulted with older blocks thrown above and over younger blocks. The ages of the rocks, crucial to these interpretations, have been worked out by the use of fossils and other means (Chapter 20).

Pulling apart forces have created the faults of the Basin and Range region that covers most of Nevada, and parts of Oregon, Idaho, California, Utah, Arizona, and New Mexico. Movement along opposing faults has formed down-dropped blocks, *grabens* (GRAH-buhnz), which are revealed as elongate basins or rift valleys today, much as the rift valleys in eastern Africa. Roughly north-south-trending mountain

Figure 9-8. Fault in marble. Jazida do Urubu, Minas Gerais, southeastern Brazil. The fault, best displayed in a dark layer to the left of the man, passes into folds toward the top of the rock face. Photograph by R.M. Wallace, U.S. Geological Survey.

ranges intervene with the grabens. For a good sense of the Basin and Range, drive U.S. Highway 50, "The Loneliest Road In America," east-west between Reno and Ely, Nevada. Your climb over the ranges and drop into the intervening grabens seems endless (Chapter 23, Central Nevada). You likely won't see the faults at the base of the mountain ranges, but geologists have documented that they exist. Faults that cut through alluvial fans (Chapter 2) verify that the faulting is new in places.

Landscapes evolve in the Basin and Range as elsewhere. In the northern part, including all of Nevada, many landscapes are in an early state of development. Relief of the ranges is high, and fault cliffs or scarps may be evident where faulting is new. In places, where erosion has persisted longer, the ranges are well cut up or dissected by streams and alluvial fans spread out into the grabens. *Playa lakes*, shallow, temporary lakes in the central parts of grabens, deposit white salts when they dry. Eventually, as portrayed in the southern Basin and Range of Arizona and New Mexico, the basins fill with sediment and the mountains are reduced to only tiny isolated remnants of their former lofty heights.

Figure 9-9. San Andreas Fault. Carrizo Plain, southwest of Bakersfield, western California. Side-slip movement along the fault has offset stream valleys. The terrain on the right side of the fault has also been raised. View is to the northwest. Photograph by R.E. Wallace, U.S. Geological Survey.

The north-south-trending Teton Range in Grand Teton National Park in northwestern Wyoming is a small but fascinating fault-block mountain range. The Range trends north-south because of a fault system along its eastern base that aligns in those directions. Pulling-apart forces have been at play here as for the Basin and Range region. The mountain block has risen while the block bearing the valley of Jackson Hole has dropped. Movement of the two blocks in relation to one another may have been 35,000 feet (10,670 meters). Imagine, a staggering 6.5 miles (10.5 kilometers)! Because of the faulting the Tetons have no foothills on their eastern side. Geologists believe that most or all of the shifting along the fault system has taken place within the last 9 million years.

Horizontal movement of broken rock masses alongside one

another, with little or no vertical motion, produces another kind of landscape. To visualize this, let's turn to a classic horizontal-slip fault that runs along most of the western side of California--the San Andreas (Figure 9-9). In actuality, the San Andreas is not a single fault but a belt of faults about 300 feet (91 meters) or more wide made up of broken and ground-up rock. Along this fault zone are straight valleys often with streams or ponds or lakes. San Andreas Lake, the namesake of the fault, is one of these lakes just south of San Francisco. In other places, linear ridges are present along the fault. Where streams flow across the fault with new movement along the fault, a distinctive pattern results. Stream channels are offset and make right-angled bends where they cross the fault.

PART 2. PRACTICAL GEOLOGY: COPING WITH GEOLOGIC HAZARDS

Chapter 10
Floods

Most of the time, streams remain within their channels. But during floods streams spill out of their channels to inundate the flat floodplains that flank them. Well before this happens, it's time to reach for high ground.

Many major cities rest beside streams: Cincinnati on the Ohio; St. Louis and New Orleans on the Mississippi; Portland, Oregon on the Columbia. And for good reason. These major streams serve as transportation avenues for ships and barges, and the flat floodplains next to them are good building sites and provide fertile growing soil. But places of high population along streams, at times, pay a price.

Kinds of Floods and Their Damage

Floods are categorized by how often they occur and their volume. A 100-year flood, of considerable volume compared to that of a 10-year flood, occurs--*on average*--every 100 years. A better way of looking at this is that a 100-year flood has a 1-in-100 chance of happening in any given year. But bear in mind that a 100-year flood may occur in two successive years or even twice in the same year.

Floods usually occur from melting snow in spring and heavy rains; a real aggravation is more-than-normal meltwater coupled with rainwater at the same time. Further, a flood in spring may be followed by another flood in summer from torrential rains.

Floods inflict damage in many ways. High water volume undercuts banks to cause buildings and piers to give way and highways and railroad embankments to be cut through. Floodwater invades buildings to destroy floors and walls, coats surfaces with mud, causes sewers to back up, and contaminates water supplies.

Cities aggravate the destructive effects of floods. Paved streets, parking lots, roads, and storm sewers increase the volume and rate at which floodwater races through a city as runoff. Sinking of the floodwater into the ground is not possible except in lawns, gardens, and in parks. A further aggravation is that bridges, docks, and buildings tend to constrict the flow of the water to heighten the level and increase the speed.

Flood Control

City planners and engineers try several approaches to control floods. One is the use of dams on main streams as well as the tributaries to trap water at times of high water levels. Dams help in some places, are impractical and too costly in others. But dams catch sediment as well as water, and eventually the reservoirs behind them fill up. Dikes, also called levees, are embankments of sediment and soil built next to stream channels to help confine floodwaters. Besides dikes constructed in cities, farmers build them to protect their farmsteads. Walls of stone or concrete on the outside of stream bends help slow erosion during floods.

All these measures can help but they may give a false sense of security. If you construct a dam or dike for a 50- or 75-year flood, what is your recourse if next year offers a 100-year flood? What can be really scary is to live in a place where the top of a dike or dam is well above the top of your dwelling.

Another, but costly, option for flood control is straightening a winding stream channel--channelization--in the vicinity of a city, or building a diversion canal around one or two sides. Either approach moves floodwater faster through a flood-prone area and doesn't let it pile up so as to diminish the effects of flooding.

Besides flood control other procedures may dampen the effects of flooding. Floodplain zoning is a major one: limiting building on floodplains that are especially prone to flooding. Such places can be set aside as parks that would not be appreciably damaged by a flood. Although we're often swamped by many kinds of insurance, flood insurance may be a necessary option. Revamping a flood-damaged dwelling can by costly. Finally, some attempt at floodproofing a dwelling may be prudent. At least taking steps to make floors and lower walls waterproof.

For nearly three decades I lived in a flood-prone city, Grand Forks, North Dakota on the Red River of the North. It lies within the flat glacial Lake Agassiz plain (Chapter 7), fertile but detrimental during times of high water. Dikes in parts of the city help contain floodwater. One April a flood was serious enough to call off classes at the University of North Dakota, and all able-bodied persons were requested to help battle the flood. I volunteered, and one night I and a shivering crew placed sand bags on top of a dike to raise it. The water level was within a foot of the top of the dike. When we thought we were gaining, rain began to fall.

During my years in Grand Forks, my dwelling was never in danger from floodwater, and I ignored flood insurance and rationalized: "If I go, the whole town goes." Five years after I left, in 1997, the whole town did go. Several feet of floodwater occupied my former dwelling. In irony, downtown buildings caught fire while water covered the streets. Firefighters had to battle water to put out the fires.

Flash Floods

Flash floods are sudden and short-lived, triggered by intense thunderstorms. Sometimes you have little time to avoid these floods. A notorious flash flood took place in Big Thompson Canyon west of Loveland in north-central Colorado in July 1976. Up to 12 inches (30 centimeters) of rain fell in the region in two days. The Big Thompson River swept through the canyon with speeds up to 15 miles (24 kilometers) per hour. Savage waters wiped out homes, cabins, motels, vehicles, bridges, a small dam, and part of U.S. 34, and moved boulders 20 feet (6 meters) across. Near the canyon's mouth, where no rain fell, the river rose 14 feet (4 meters) in a few minutes. One hundred thirty nine people died in the flood, many while trying to outrun the floodwaters by driving down the canyon. Road signs still warn of possible flash floods in the canyon.

Flash floods must also be especially watched for in *slot canyons*, narrow, straight-sided canyons often in resistant sandstone. Many occur in southern Utah. Beautiful to photograph, and enticed by the yellow, oranges, and reds, you must also take heed of the dangers upon entering slot canyons.

To enter some slot canyons from their tops can be hazardous in itself when their bottoms are unknown. In places, their tortuous sides may be just barely wide enough to squeeze through. While the channels of many slot canyons are normally dry, some contain deep pools that must be waded through. Large boulders and tree trunks may be real obstacles to navigate around.

Never enter a slot canyon with the slightest chance of rain. Even if local rain is not imminent, thunderstorms several miles away may place you in peril. Imagine the consequences of being trapped in a slot canyon during a flash flood. You might have traveled too far to escape at your entry point, and your chances of finding a new egress within moments are meager. Obstacles in the channel hinder your hurried, anxious movements. Just before the noisy water slams your slot, you grope for higher ground as the sandstone bruises your limbs and body. No need for me to be melodramatic, perhaps, but a slot canyon adventure can be a risky business.

Personal Steps to Take Against Floods

Some steps to take against floods are as follows:

1. Examine a flood-risk map for your region or city, if possible before you take up residence. Obtain this map from a state or provincial geological survey or the city.

2. If you have the choice, avoid living on a floodplain.

3. Be a geologically-informed voter. Vote for the flood-control measure that makes most geological sense, be it dike, channelization, dam, or diversion canal.

4. If you already live in a flood-prone area, consider buying flood insurance.

5. In regions susceptible to flash floods, keep a wary eye on the occasional thunderstorm that leads to a cloudburst. Don't enter a narrow canyon if you know of an impending rainstorm even if miles away.

6. For more information on floods, contact this U.S. Geological Survey's web site: www.usgs.gov/themes/flood.html.

Chapter 11
Earthquakes

Earthquakes are vibrations or tremors on Earth that result from the sudden release of energy. Most earthquakes occur when Earth forces subject rocks to so much stress that they break along faults. Some earthquakes are set off by movement of molten rock at depth prior to a volcanic eruption. The source of an earthquake is at some depth, and the point on the surface above that source is the *epicenter* (from the Latin *epi,* over and *centrum*, center).

Seismic Waves

Three kinds of *seismic waves* (from the Greek *seismos*, shock or earthquake) streak from the source of an earthquake. Two types travel through Earth's interior as they spread outward from the source in all directions. The other type moves on the surface away from the epicenter like water waves rippling from a pebble thrown into a pond. *Seismographs* are instruments that pick up and record seismic waves as wiggly lines on a paper records.

Because the three kinds of seismic waves travel at different speeds, seismographs can locate earthquakes. The times of arrival of seismic waves at a seismograph relate to distances traveled. Seismologists--earthquake specialists--draw these distances as arcs of circles from three seismographs. The point where the arcs intersect is the epicenter of an earthquake.

Measuring An Earthquake's Strength

An indication of an earthquake's strength is its *magnitude*, the amount of energy released, given as a number. Most often an earthquake receives a number from the *Richter scale*, 0 to 8.6--the greatest on record. Any earthquake ranked as 7 or greater is a major earthquake. Measuring the size of seismic waves from a seismograph record gives the magnitude. Bear in mind that each whole number step on the Richter scale is a 10 times increase in magnitude. Vibrations of an earthquake with a magnitude of 3, therefore, are 10 times greater than those of an earthquake with a magnitude of 2. A tenfold increase in the size of earthquake vibrations translates to about a 30 times increase in the energy released.

Seismologists also use "moment magnitude" to reflect the energy released during an earthquake. This is based on the surface area of rupture and the amount of rock shifting along a fault. Magnitude on the moment scale may be higher or lower than Richter magnitude, and may also be higher than 9. The news media don't always distinguish whether their numbers are for the Richter or the moment scale.

Effects of Earthquakes

Earthquakes may affect a region in many ways. The most obvious effect is ground motion, which if strong, may be visible. Slight motion may crack walls or windows, intense motion may sever roads and bridges and topple tall buildings. In cities, people are injured or killed mostly by falling debris from buildings. So, unless on an upper-story floor, you are usually safer inside a building than outside of one. Strict building codes that stress good building design help lessen damage from earthquakes.

Smaller *aftershocks* may follow the main earthquake, and can also produce physical damage as well as added psychological distress after having experienced a primary tremor. Following the southern Alaska earthquake near Anchorage in 1964, with a highly major Richter magnitude of 8.6, several thousand aftershocks peppered the region from March to December.

Figure 11-1. Surface faulting associated with March 27, 1964 Alaskan earthquake. Anchorage, Alaska. One school building has been broken in two, another wrinkled. The ground in the central part of the photograph has dropped about 12 feet (4 meters). Photograph by W.R. Hansen, U.S. Geological Survey.

The land surface may be displaced horizontally or vertically as rock masses shift along faults. A fault may be evidenced by a tear on the surface if by horizontal movement or by a low cliff if the movement was near vertical. Horizontal displacement tied to the 1906 San Francisco earthquake (8.2 on the Richter scale) offset roads, fences, and buildings up to 23 feet (7 m). In rare cases, the ground may open slightly along a fault. During the 1964 Alaskan earthquake, parts of the coastal ocean floor was raised above sea level and other areas sank. Surface displacement along faults ripped apart buildings

Figure 11-2. Cracking of paved road associated with May 31, 1970 Peru earthquake. Western Chimbote, Peru. Liquefaction of water-saturated beach sand beneath the road caused its fracturing.

(Figure 11-1).

Earthquakes often set off landslides. The Madison Canyon rockslide west of Yellowstone National Park (Chapter 8), triggered by an earthquake with a Richter magnitude of 7.7, is a classic example. In 1970, a Peruvian earthquake (7.8 on the Richter scale) in the Andes Mountains triggered thousands of landslides and entombed more than 17,000 people.

Another geologically-derived effect is *liquefaction* (Figure 11-2), whereby water-saturated soil and sediment virtually turns into a liquid from an earthquake's agitation. Liquefaction may occur within minutes of an earthquake to cause buildings to sink, tilt, or topple, and underground tanks to float to the surface. Liquefaction caused considerable damage in the 1989 Loma Prieta earthquake (7.0 on the Richter scale) near Santa Cruz, California, and contributed to destruction in the 1964 Alaskan and 1906 San Francisco earthquakes.

Earthquakes are one means to trigger tsunami, giant sea waves. I devote Chapter 12 to these geologic hazards.

Last to mention, but not to belittle its effect, is fire. In the 1906 San Francisco earthquake, fire caused most of the damage to the city. Broken gas mains and fallen electrical wires lit the fires and broken water mains made dousing the fires difficult.

Where Earthquakes Occur

The majority of earthquakes are found in two narrow belts, one that circumscribes the Pacific Ocean, the other passing from the Mediterranean Sea to the Himalayas and the East Indies. This occurrence coincides with the distribution of composite volcanoes (Chapter 5). I mention in Chapter 20 the theoretical basis for this agreement in occurrence of earthquakes and volcanoes.

Earthquakes can also be grouped by their occurrence with depth. Eighty–five percent have a shallow source, 0 to 44 miles (70 km); 12 percent are at intermediate depths, 44 miles to 218 miles (350 km); and 3 percent are deep–seated, 218 miles to 416 miles (670 km). Earthquakes with deeper sources inflict less damage because deep rocks yield by gradually bending, not breaking abruptly to release vibrations. The 6.8-magnitude Nisqually earthquake of February 2001, epicentered between Seattle and Olympia, Washington, caused relatively little damage because of its deep source 32 miles (52 km) below the surface.

In the United States, most of the larger (magnitude 4.5 and greater) and most damaging earthquakes occur in the western states, especially southern Alaska, California, Nevada, Utah, Idaho, Montana, and Washington, and New Mexico. Although less common east of the Rockies, earthquakes can be expected to inflict moderate damage in several places: southeastern Nebraska and northeastern Kansas; a region involving parts of Missouri, Arkansas, Tennessee, Kentucky, and Illinois; South Carolina and part of Georgia; and central New England.

Predicting Earthquakes

Predicting earthquakes is more of an art than a science. Several clues, however, provide insight into an earthquake's possible happening. An easy first step is to compile the history of earthquakes in a region, and use this history to assess the probability of earthquakes in the future. Along some faults are inactive sections where earthquakes have not occurred for a long time. These gaps in earthquake occurrence are places where the next earthquakes are apt to strike--perhaps major ones.

Seismologists use sensitive instruments to measure tiny amounts of tilting, bowing, uplift, or sinking of a land surface as rocks swell before rupture. These movements may suggest an oncoming temblor. In like manner is the measurement of minute cracks in rocks along a fault, which may set off micro-tremors. The opening of these cracks changes the amount of pore space in rocks and may cause the water level in wells to rise or fall preceding an earthquake. Such cracks may also increase the amount of radon gas (Chapter 16) released from wells which helps prediction.

When an earthquake is close to imminent, other approaches may allow short-term prediction. *Foreshocks*, which may precede a main earthquake by many hours, should definitely be considered important. Hundreds of foreshocks for five hours preceded a magnitude-7.3 earthquake near Haicheng in northeastern China in 1975. This warning allowed ample time for millions of people to evacuate even though the earthquake demolished half of the buildings in the city.

Speaking of the Chinese, they tend to rely with seriousness on animal behavior. Prior to the Haicheng earthquake, snakes came out of hibernation in winter to freeze on the snow. Agitated rats showed no fear of humans. And pigs chewed off their tails and ate them.

Personal Steps to Take Against Earthquakes

1. Be aware that weaker foreshocks may precede and aftershocks may follow a strong earthquake. Pay special attention to foreshocks as warnings of something stronger. Avoid upper stories of tall buildings or possible falling debris from them if you sense even the slightest of tremors. If near one- or two-story buildings, a relatively safer place may be in the streets.

2. Stay away from landslide-prone places if you experience foreshocks.

3. Keep in mind any erratic animal behavior that may relate to a forthcoming earthquake.

4. Request an earthquake-risk map from a state geological survey or a provincial geological survey. Use it to evaluate the relative chance of an earthquake happening where you live or where you intend to live.

5. If the earthquake risk is high for item 4 in your area, you may wish to purchase earthquake insurance.

6. Determine the geological material beneath your dwelling or intended dwelling. You can do this from a geological map published by a geological survey mentioned in item 4. Be aware that buildings built on soft sediment are more susceptible to damage than those constructed on solid rock. Sort of like sitting on Jell-0 versus concrete.

7. Be aware that in cities the greatest damage from earthquakes may result from fire brought about by ruptured gas mains and fallen electrical wires.

8. Consider the following if they apply to you:

     a. Stabilize water heaters with special leg supports.

     b. Brace the foundation of a mobile home.

     c. Install a seismic gas shutoff valve.

9. To stay attuned to earthquakes nation-wide as well as world-wide, contact the National Earthquake Information Center at http://neic.usgs.gov/. You can also order earthquake-risk maps through this web site.

Chapter 12
Tsunami

Tsunami (soo-NAH-mee) are hazardous water waves. Often misnamed "tidal waves," they have nothing to do with tides. Tsunami are generated by any disturbance associated with earthquakes, volcanic eruptions, underwater landslides, or the impact of a meteorite with the ocean. Although most often in oceans, these destructive waves may also be expected in lakes and reservoirs.

Tsunami is a Japanese word that means harbor (*tsun*) wave (*ami*). Both the singular and plural of the word in Japanese are spelled the same.

Features of Tsunami

Because energy is transferred from a seafloor disturbance, tsunami wave motion involves the entire depth of the ocean. Tsunami, therefore, differ from ordinary water waves in that they are higher (near shore), move faster, and have a longer wave length--mentioned in Chapter 3 as the distance from the high point of one wave to the high point of another. Near shore, tsunami may reach heights of 50 to 100 feet (15 to 30 meters) or more, although in mid-ocean they may be imperceptible. Their speeds may reach that of a commercial jetliner. And their wave lengths may be as long as 310 miles (500 kilometers)!

A tsunami often runs up on shore at a height twice or more than the height of the wave as it approaches the shore. The greatest run-up height against a shore recorded is 1,719 feet (524 meters) above sea level at Lituya Bay, Alaska in 1958 (Figure 12-1).

Occurrence of Tsunami

Most tsunami occur in the Pacific Ocean and in the region of the East Indies. In the Pacific most are generated in Japan and South America by earthquakes. Specific highly tsunami-prone places include southwestern Chile, the Chile-Peru border region, Hawaii, and northern California. The largest tsunami of the 20th century occurred in Moro Gulf, Philippines in 1976 killing 8,000 people. Only the South Atlantic doesn't seem to have tsunami and the North Atlantic is almost free of them.

Causes of Tsunami and Examples

Baymouth Bar

Figure 12-1. Tsunami damage of forest. Lituya Bay, Glacier Bay National Park and Preserve, southeastern Alaska. Lituya Bay opens into the Gulf of Alaska. A rockslide at the head of the bay generated a tsunami that destroyed the forest around the bay. A baymouth bar nearly isolates the bay. Photograph by D.J. Miller, U.S. Geological Survey.

Earthquakes are the most common cause of tsunami, and two-thirds of damaging tsunami associate with earthquakes having a magnitude of 7.5 or more. Many earthquakes, however, don't produce tsunami. Keep in mind that earthquakes on land, as well as on sea bottoms, can produce tsunami.

A major earthquake triggered many tsunami along the southern Chilean coast in May 1960. These tsunami reached 26 feet (8 meters) high and speeds of 124 miles per hour (200 kilometers per hour). Run-up of tsunami onto the coast near the source area reached 82 feet (25 meters) above sea level. An estimated 5,000 to 10,000 Chileans lost their lives.

Within 24 hours a series of tsunami swept across the Pacific. On islands with gradual sea-bottom slopes and bays, tsunami wave heights may be amplified several times.   Hawaii was hit particularly hard, just less than 15 hours after the Chilean trigger. Although 6,200 miles (10,000 kilometers) from southern Chile, 61 people lost their

Figure 12-2. Tsunami damage, from Alaskan earthquake of March 27, 1964. Valdez, Alaska. Associated damage resulted from the earthquake itself and submarine landslides. The town and its docks were destroyed. Photograph by U.S. Geological Survey.

lives at Hilo. Up to 20-foot (6-meter) tsunami swept 10-ton (9-metric ton) vehicles away and bent parking meters flat. Both earth tremors and landslides produced tsunamis associated with the main southern Alaska earthquake in 1964. A wave at Whittier reached 105 feet (32 meters) above sea level. Another at Valdez (Figure 12-2) lifted driftwood170 feet (52 meters) above sea level and silt and clay 49 feet (15 meters) higher. Tsunami accounted for 106 deaths.

The main tsunami spread southward into the Pacific within 25 minutes of the main earthquake. At many places, the first wave caused a rapid rise in sea level, followed by a lowering of sea level and finally a bigger advancing wave. This was the largest historical tsunami disaster to strike the West Coast of the U.S. with wave heights to 15 feet (4.5 meters). Hardest hit was Crescent City, California with 11 people killed. This tsunami-prone city has a sea-bottom topography that tends to concentrate a wave's energy. Before the fourth wave hit, water withdrew from the inner harbor, and left it dry.

The first wave struck Hilo, Hawaii 1.3 hours after its strike at Crescent City. Little damage occurred this time because of the long warning time, a low run-up onto shore of only 10 feet (3 m) high, and because much of the area smashed by the 1960 Chilean tsunami had not been resettled.

Underwater landslides may occur on the steep slopes of volcanoes, the sides of ocean trenches, seamounts, and atolls. Some produce tsunami greater than those generated by earthquakes. Earthquakes trigger most landslides.

An earthquake set off a landslide in the Grand Banks area of eastern Canada. Two and one-half hours after the earthquake, a 10-foot-high (3-meter-high) tsunami burst onto the south coast of Newfoundland at 87 miles per hour (140 kilometers per hour), and ran up onto shore as much as 88 feet (27 meters). Twenty-eight people lost their lives in Newfoundland.

Another bit of notoriety is tied to this landslide. The landslide transformed into a turbidity current--a rapidly-moving water-sediment slurry--that cut 12 telephone cables on the sea bottom between Canada and Europe in 11 hours. Its calculated speed reached 66 feet per second (20 meters per second).

Although the North Atlantic has been relatively free of tsunami, people living around this ocean should not be complacent. Scientists especially fear a potential "mega-tsunami" triggered in the Canary Islands off the northwestern coast of Africa. One island, La Palma, consists mainly of volcanic rubble. If a gigantic block on its western flank should drop catastrophically into the sea during an eruption, a gigantic tsunami might ensue. Studies show that a landslide-induced tsunami more than 2,130 feet (650 meters) high would wipe out cities along the U.S. East Coast.

Volcanoes generate tsunami in various ways. Shallow, relatively quiet underwater eruptions can disturb the water to form tsunami. More violent tsunami result from volcanic explosions where molten rock material bursts into seawater. Flows of volcanic debris can cascade down volcano slopes to generate tsunami. And, as mentioned above, earthquake-triggered landslides race down steep slopes of volcanoes.

One of the largest volcanic explosions took place on the Indonesian island of Krakatoa (kra-cuh-TOE-uh) between Java and Sumatra in 1883. The loudest of several blasts was recorded 3,000 miles (4,800 kilometers) away. The largest tsunami, which killed thousands of people, reached 130 feet (40 meters) high. Coral blocks up to 600 tons (544 metric tons) were shoved onshore.

Tsunami caused by impact of meteorites or comets with the ocean is not backed by any historical record. But prehistoric evidence does exist. One of the clearest is a buried impact crater on the Yucatan Peninsula at Chicxulub, Mexico. Sedimentary deposits laid down by tsunami have been recognized on land around the Gulf of Mexico. The impact event dates at 65 million years ago, and some believe it brought about the extinction of the dinosaurs (Chapter 21).

Tsunami Warning Systems

Two centers cover comprehensive warning systems for the Pacific: the Pacific Tsunami Warning Center (PTWC) maintained by the U.S. National Weather Service and the International Tsunami Information Center (ITIC) run by UNESCO. The National Oceanic and Atmospheric Administration (NOAA) and the U.S. Coast Guard broadcast tsunami warnings, and you can subscribe to e-mail warnings from *WWW.TSUNAMI@ITIC.NOAA.GOV.*

Personal Steps to Take Against Tsunami

1. If you have considerable warning, head inland for the highest ground you can reach. Don't rush to return seaside: Remember that many tsunami are possible in one event, and that later tsunami may be stronger than earlier ones.

2. If you feel an earthquake along a coast, head inland for higher ground immediately. Don't wait for a tsunami warning.

3. Take heed from a rapid withdrawal of water from the shore. This is a clear indication of an impending tsunami, but all do not give this warning.

4. In a city, seek haven at an upper floor of a tall, reinforced-concrete building.

5. Avoid standing on cliffs to witness forecast tsunami, which may reach 100 feet (30 meters) or more above sea level.

6. Avoid harbors, bays, gullies, deep tidal rivers and estuaries, and headlands. These places tend to amplify tsunami.

7. If you have a choice in your quest for safe places, choose forested land over cleared land, pastured floodplains, and paved urban areas over which tsunami travel with greater ease. In rural areas, you may be forced to climb a tree and lash yourself to it.

8. Be aware of tsunami warnings from NOAA and U.S. Coast Guard broadcasts.

Additional Reading

Bryant, Edward. *Tsunami: The Underwater Hazard.* Cambridge University Press, 2001.

Chapter 13
Volcanic Eruptions

Volcanoes may be grouped by their general behavior. Active volcanoes have a historical record of eruption, and may be expected to continue this trend. Dormant volcanoes don't have a historical record but appear fresh, little eroded. These "sleeping giants" can be especially dangerous. Mt. Rainier in southwestern Washington is a dormant volcano that must be watched closely, along with other peaks in the Cascade Range. Extinct volcanoes have no historical record but a prehistoric record, appear notably eroded, and may be considered "dead." But the possibility exists that a so-called extinct volcano may be actually dormant.

Volcanic eruptions are of two types: quiet and explosive. Quiet eruptions involve molten rock material that issues to the surface with relative gentleness because it contains little gas. Such eruptions produce shield volcanoes (Chapter 5) and lava flows that escape through cracks in rocks called fissures. Explosive eruptions happen when molten rock material releases gas suddenly as by the quick opening of a plugged vent or by making sudden contact with water. Cinder cones and composite volcanoes develop from explosive eruptions.

Example of Quiet Eruptions

Kilauea (kih-low-AY-uh) is one of five volcanoes that make up the island of Hawaii and is the most active. Most outpourings of lava reach the surface quietly but explosive activity occurs. Lava fountains may spurt hundreds of feet (hundreds of meters) high upon release of gases, and cinder cones may build up. Lava from the fountains falls into lava ponds or lakes and feeds flows, some of which may issue from fissures along the flanks of the volcano and wind their way to the sea. Temperatures of the flows may reach 2,200 degrees Fahrenheit (1,200 degrees Centigrade).

Two main tools help forecast Kilauea's eruptions--that they will likely occur but not necessarily when. Seismographs (Chapter 11) register tremors that indicate shifting of molten rock material beneath the surface. Sometimes swarms of thousands of micro-tremors may precede an eruption. Tiltmeters, like giant carpenter's levels, measure the tilting outward of the flanks of the volcano as well as their bulging upward. The swelling of the flanks is often a good indicator of an imminent eruption.

Kilauea's mostly quiet eruptions of largely slow-moving lava rarely injure anyone or take their lives. Many buildings, however, are often destroyed and sometimes entire villages.

Maybe not apparent at first thought, Kilauea's eruptions benefit as well as destroy. Flows that travel to the sea allow the island of Hawaii to grow. And, for that matter, volcanic eruptions have built up the entire Hawaiian Island chain above the floor of the Pacific Ocean over a period of millions of years. Through weathering--the breakup and chemical interaction of rocks with air and water--lava and volcanic ash decay into fertile soil over time.

Examples of Explosive Eruptions

The largest volcanic catastrophe of the twentieth century destroyed St. Pierre, a port city on the Caribbean island of Martinique within the Lesser Antilles. The main eruption of Mt. Pelée (puh–LAY), 4,582 feet (1,397 meters) above sea level and 5 miles (8 kilometers) from the city's center, created a devastating *pyroclastic flow* of hot gases, volcanic ash, and rock fragments that swept down the flanks of the volcano. Pyroclastic, from the Greek *pyros*, fire and *klastos*, broken, is a technical term with which many of those who live in the shadow of explosive eruptions are familiar. Other names for pyroclastic flow are "ash flow" and "glowing avalanche." "Glowing" refers to the high temperatures of the gases that cause a pyroclastic flow to glow at night. A pyroclastic flow actually consists of a ground-hugging avalanche of hot gases and rock fragments topped with an overriding turbulent dust cloud.

The Mt. Pelée pyroclastic flow, racing at an estimated 60 miles per hour (96 kilometers per hour), swept away buildings and people in its path, pushed over stone buildings, overturned and burned boats in the harbor. An estimated 29,000 people died. A single survivor was a prisoner in a basement cell. Deaths resulted from falling buildings, flying rocks, hot gases that burned lungs and boiled body fluids, and inhaled ash that clogged lungs and noses.

Because many ravines and valleys separated the volcano from the city, people thought they were safe from an eruption. This assumption was correct if only lava flows were threatening. But even geologists were unfamiliar with pyroclastic flows at the time, and their capability to flash over great distances and obstacles at incredible speed.

Mt. Pelée erupted again in 1929 but caused less destruction. Further eruptions are definitely possible. Hot springs bubble nearby.

Some good has come from this eruption. The French government has quarried millions of tons of pyroclastic debris to upgrade and resurface roads on Martinique.

Mt. Vesuvius is a Mediterranean volcano with a long explosive history. It rises 4,200 feet (1,280 meters) above the Bay of Naples in southern Italy, 10 miles (16 kilometers) from Naples' city center. Vesuvius has erupted more than 50 times within the last 2,000 years. Its most infamous eruption took place in A.D. 79.

For 16 years earthquakes preceded the A.D. 79 eruption, which generated pyroclastic flows and volcanic ash fallouts. Mud, volcanic ash, and rock fragments buried the Roman cities of Pompeii and Herculaneum near Naples, along with *pumice* (PUHM-uhs), a glassy froth, so porous that it floats on water. These cities were forgotten until more than 16 centuries later when they were excavated. Most of Pompeii's 20,000 citizens succumbed to burial, and many are believed to have been asphyxiated by sulphurous fumes. Carbonized bodies decomposed, leaving cavities in the ash. Archaeologists poured plaster of Paris into the body cavities to produce striking casts of the fallen inhabitants in various agonized poses.

Vesuvius' eruptions were almost continuous from 1631 to 1944, and the volcano must be considered "active" today. Naples is well within the reach of pyroclastic flows.

Mt. St. Helens, thought of as "America's Vesuvius" by some, is one of 15 major composite volcanoes aligned north-south within the Cascade Range from southern British Columbia to northern California. As mentioned at the beginning of the chapter, all of these peaks should be considered dormant with Mt. Rainier next to Mt. St. Helens in the list of those potentially most hazardous. Lassen Peak in northern California erupted during 1914 to 1917, second only to Mt. St. Helens in recent activity.

St. Helens, dormant for 123 years, awakened in March 1980. Thousands of small earthquakes preceded steam and ash eruptions, all clues that molten rock material was moving toward the surface. The volcano began to swell, and the northern flank bulged as much as 5 feet (1.5 meters) per day. The U.S. Geological Survey feared that the bulging would set off a mammoth landslide which happened as predicted. A fairly strong earthquake shook the bulge loose and the ensuing landslide removed the lid that sealed in molten rock. On May 18, within seconds, the sudden release of trapped gases caused a massive explosion. You might compare this with shaking a container of warm soda and popping it open, in this case the release of carbon dioxide causing the "explosion." The blast has been estimated as having had 500 times the force of the atomic bomb explosion at Hiroshima during World War II.

Catastrophic   effects   were   striking.   The   landslide   and explosion removed a fourth of the volcano. Pyroclastic flows, at speeds to 80 miles per hour (130 kilometers per hour) and temperatures of about 900 degrees Fahrenheit (500 degrees Centigrade), killed the forest with their hot breath 15 miles (24 kilometers) away. Mudflows, mostly hot, from rain and melting snow and glacier ice, raced down stream valleys many miles (kilometers) to kill people and destroy homes, vehicles, bridges, and timber (Figure 8-3). Fallout of ash, like gray snow, made breathing difficult. Volcanic particles were sucked into automobile engines and destroyed them. The ash cloud drifted over Cheyenne, Wyoming the day after the eruption and encircled Earth by June 5. The volcano erupted twice more in July and August before it became quiet.

Sixty-three people lost their lives from St. Helens' explosion and related effects, including a geologist stationed at an observation post 5 miles (8 kilometers) northwest of the crater. Many more would likely have perished if not for the warnings of the U.S. Geological Survey to public officials. As important was the resistance to pressure by these officials from the public to gain access to lands closed because of the perceived threat of eruption.

Are further eruptions possible? Very likely, considering St. Helen's fiery history. Evidence exists for more than 20 eruptions in the past 4,500 years. Periods of dormancy have lasted from tens of years to 500 years or more. Eruption comes down to when not if.

Personal Steps to Take Against Volcanic Eruptions

1. Be alert for earthquakes near active or dormant volcanoes. They may be precursors to impending eruptions.

2. Take heed of possible eruption warnings from federal or state agencies. Resist complacency toward these warnings if some don't materialize.

3. Remind yourself of the differences between quiet and explosive eruptions, and prepare accordingly.

4. Be mindful of the added danger of high-speed pyroclastic flows and volcanic mudflows that may travel many miles from a volcano. Pyroclastic flows especially are not necessarily deterred by intervening valleys or ridges.

5. Realize that volcanic ash can plug automobile engines and render your escape from an eruption difficult if not impossible.

Additional Reading

De Boer, Jelle Zeilinga, and Donald Theodore Sanders. *Volcanoes in Human History: The Far-Reaching Effects of Major Eruptions.* Princeton University Press, 2001.

Chapter 14
Subsidence

Subsidence, the gradual or sudden down-sinking of rock and rock debris, differs from landslides in that the gravity-forced movement is mostly vertical rather than down a slope. In preceding chapters, I've already mentioned the natural collapse of caves and caverns to form sinkholes, the collapse of volcanic craters to form calderas, the melting of isolated ice blocks to form kettles, and the collapse of lava tubes. Two other nature-induced kinds of subsidence happen: when permafrost thaws and when coal seams catch fire naturally and roofs of burned-out cavities collapse into the space left by the coal.

In 45 states of the U.S., an area the size of New Hampshire and Vermont combined is directly affected by subsidence. Both human-caused and natural subsidence are involved.

Examples of Human-caused Subsidence

More than 80 percent of subsidence in the U.S. is human-caused, primarily by the over-withdrawal of groundwater. The San Joaquin Valley in central California (Figure 23-2), roughly between San Francisco and Bakersfield, has been subjected to such treatment. In fact, the Valley has been labeled as the "Largest human alteration of the Earth's surface." The dry Valley is rich in agricultural production, and its demand for groundwater for irrigation has been great. The greatest subsidence took place in the central part of the Valley near Mendota: 30 feet (9 meters) in 52 years (1925-1977).

When groundwater is withdrawn faster than it's replaced--recharged--by trickling into the ground from stream channels and precipitation, water drains from the pore spaces and they collapse. The aquifers, therefore, undergo compaction and the ground above sinks or subsides. In the San Joaquin Valley since 1974, subsidence has largely slowed or stopped with reduced pumping of groundwater for irrigation and the use of more surface water. But during later drought periods and greater dependence on groundwater, subsidence has again raised its ugly head. The relationship with over-withdrawal of groundwater and subsidence in the Valley is established.

To quench its thirst, Las Vegas sinks. Excessive "mining" of groundwater has caused the city to subside nearly 6 feet (1.8 meters) between 1960 and 1990. As for the San Joaquin Valley, over-withdrawal of groundwater has caused the aquifers to compact and the ground surface to sink. Uneven compaction and settling has created fissures which damage homes, roads, railways, curbs, and swimming pools; and rupture sewage, water, and gas lines. Through erosion, fissures transform into gullies. An interesting twist is that fissures often occur along pre-existing faults, which, themselves, are often barriers to groundwater flow.

In pre-settlement days, Las Vegas--Spanish for "the meadow"--was a desert oasis with welcomed springs and creeks. But its present problems initiated from an early demand for more water. The first well in 1907 was increased to nearly 125 by 1912. By 1955, more water was withdrawn than recharged, and groundwater levels dropped more than 90 feet (27 meters) during 1944-1963. Surveyors realized the city was sinking by the 1940s.

Water managers try to curb Las Vegas' subsidence by replenishing groundwater artificially. They pump treated water from nearby Lake Mead during the cooler, lower-demand months to raise groundwater levels. This practice raises groundwater levels, but overall discharge from the wells still exceeds natural recharge of the aquifer system. Downtown Las Vegas no longer subsides but the northwest part of the city still sinks at 1 to 1.2 inches (2.5 to 3 centimeters) per year. The added quench for Las Vegas' thirst as it grows threatens more subsidence.

Example of Natural Subsidence Aggravated By Human Activity

In Chapter 6, I mentioned that sinkholes may develop in rock gypsum and rock salt, but are most common in limestone (and in the related carbonate rock dolostone). Subsidence in rock gypsum and rock salt is especially prevalent in western Texas, southeastern New Mexico, western Oklahoma, and western Kansas. Sinkholes in carbonate rocks can be found in about 40 percent of the contiguous U.S., especially east of the longitude of Tulsa, Oklahoma. Let's take a closer look at sinkhole formation, complicated by human activities, in west-central Florida.

Carbonate rocks underlie most of Florida and make the state prone to sinkhole formation. Sand and clay, of varying thickness, cover the carbonate rocks. Moving groundwater dissolves cavities in the rocks. Where the sediment cover is very thin or sandy, sinkholes develop slowly over years or centuries; where the cover is clayey, sinkholes may form by collapse rapidly over a period of hours.

In west-central Florida the effects of subsidence and the formation of sinkholes is particularly prevalent. Rapidly-formed sinkholes damage buildings and roads. Plugged sinkholes form numerous ponds, lakes, and wetlands, and cause flooding in places.

Excessive groundwater pumping contributes to sinkhole

formation as it does for subsidence in the San Joaquin Valley and Las Vegas. Groundwater levels drop to stress supporting materials and collapse occurs, usually abruptly, and revealed by sinkholes. These sinkholes may drain streams, lakes, and wetlands and cause surface water to contaminate groundwater. Fruit growers pump much warm groundwater during extended freezes to spray plants with a protective coating of ice. The longer the freeze, the more prolonged the pumping--with notable lowering of groundwater levels. Whether for irrigation or crop-freeze protection, increased pumping of groundwater is highly correlated with the formation of new sinkholes.

A dramatic sinkhole incident took place near the west Florida coast at the common border of Pasco and Hernando Counties in 1998. An irrigation well drilled through 140 feet (43 meters) of limestone, followed by a cavity at 148 to 160 feet (45 to 49 meters). Shortly after preliminary pumping and flushing of the well, two sinkholes formed near the well. With continued pumping, hundreds of sinkholes, less than a foot (0.3 meter) to 150 feet (46 meters) across, appeared within a 6-hour period in a 20-acre (8-hectare) forest. The sinkholes uprooted numerous trees.

You might surmise that the best way to avert human-induced sinkholes in west-central Florida is to control groundwater levels. This can be accomplished by restraint in groundwater pumping or by injecting water back into wells during low-demand times.

Human-induced subsidence is not only caused by over-pumping of groundwater but also of oil. At the Wilmington oil field near Long Beach, California the land has subsided 30 feet (9 meters) in 50 years from over-pumping of oil. That's bad enough, but the ensuing subsidence has placed the area in danger of another problem--flooding.

Twenty percent of land subsidence in the U.S. results from underground mining, chiefly that for coal. With underground mining comes the potential for collapse of mine roofs. Collapse produces pits, some of which resemble sinkholes. Often, though, these pits are elongate rather than circular and aligned with one another (Figure 14-1).

Personal Steps to Take Against Subsidence

1. Avoid living where ground fissures--telltale signs of subsidence--are apparent. If you already own a home where fissures are obvious, consider purchasing subsidence insurance.

2. Support and vote for officials who attempt to control groundwater levels.

3. Check your area for sinkholes on a detailed topographic

map. Seek out a possible sinkhole probability map.
    4. Avoid areas of underground mining, especially if collapse pits exist.
    5. In areas of hot springs, such as Yellowstone National Park, stay on boardwalks. Overhanging fringing crusts of thermal pools tend to give way if you step on them. People have died upon collapse of the fringing crusts and falling into scalding water.

Figure 14-1. Subsidence pits and troughs by roof collapse of underground coal mines. North of Sheridan, north-central Wyoming. The mines operated from the 1890s to the early 1920s. Overburden varies from 16 feet (5 meters) to 148 feet (45 meters).

Additional Reading

Galloway, Devin, David R. Jones, and S.E. Ingebritsen, eds. *Land Subsidence in the United States*. U.S. Geological Survey Circular 1182, 1999.

Chapter 15
Landslides and Their Effects on Humans

In Chapter 8, I discussed the kinds of landslides and the landscapes that result from them. Here, we'll consider what encourages landslides to happen and an in-depth look at snow and ice avalanches--wintry landslides.

What Helps Landslides Happen

Several conditions allow landslides to take place. Water from rain or melting snow saturates rock debris, makes it heavier and more likely to move downslope. Water also lubricates rock debris, especially the muddier kind, and reduces the sticking together of rock particles. Odd, perhaps, a little water can also prevent landslides. Where pore space is only partly filled by water, particles stick to one another by surface tension. But where pore space is completely filled with water, the rock debris-water mixture cruises with ease.

Steep slopes in themselves already bring about the potential for rock material to move, but over-steepening of slopes by undercutting considerably encourages landslides to happen. Undercutting takes place by streams along the base of valley slopes, by valley glaciers, and by waves along seacoasts with steep cliffs. Construction contractors may sometimes over-steepen road cuts or building sites.

Strong vibrations may set off landslides depending on the vulnerability of a landslide-prone slope. Among the possibilities are earthquakes, dynamite blasting, and sonic booms.

A lack or scarcity of plant cover boosts the likelihood of landsliding. Few plant roots, especially those that don't penetrate at depth, don't offer a useful anchoring effect.

A geological condition often overlooked is the orientation of planes of weakness in rocks relative to the direction of slopes. If cracks in rocks or rock layers lie parallel to slopes, rock masses have a good chance of cascading in a massive slide.

Examples of Disastrous Landslides

Earthquakes notably trigger landslides. The three landslide examples cited here share this common trigger.

In January 2001 a magnitude-7.6 earthquake jolted El Salvador. Soil and rock debris, along with part of the forest, slid down a ridge to wipe out part of Los Colinas, a San Salvador suburb. Some 450 people died and 1,200 were missing. Other smaller landslides killed fewer people and blocked roads. Deforestation did not contribute to the slide because tree roots did not extend to the depths at which underlying volcanic debris was shaken loose by the earthquake.

Another magnitude-7.6 earthquake, which destroyed buildings and killed 1,800 people in Taipei, Taiwan in September 1999, set off some 7,000 landslides in the Central Mountain Range of the country. More than 50 people died from two of the landslides. Steep valley slopes and highway cuts were most susceptible to landsliding. One landslide dammed up the Ching-Shui River to instantly create a sizable lake. Boat concessionaires quickly secured access for sightseeing and fishing.

A third example takes us to the Andes Mountains of northern Peru, where, in May 1970, the entire town of Yungay was wiped out. Although centered offshore 62 miles (100 kilometers) from Yungay, an earthquake dislodged a slab of glacier ice from Nevado Huascarán, a high peak above where Yungay once existed. The down-rushing ice broke off rock debris and scooped out small lakes to form a massive wall the size of a 10-story building. As expected, the noise from the slide was thunderous. The main mass of muddy debris leapt down a steep valley to destroy the village of Ranrahica and bury 1,800 people. A smaller part of the muddy debris mass completely entombed the town of Yungay with an estimated 17,000 inhabitants and visitors. Only the top of the church and the tops of palm trees were evident in the aftermath. The cemetery was spared because of its position on higher ground.

Avalanches

Snow and ice avalanches are basically landslides. In Chapter 8, I likened avalanches to debris slides. Although snow or ice are the main ingredients of avalanches, rock or rock debris may be appreciable as well. And, ice is a rock, and loose snow behaves in a similar fashion to that of sand, a rock when lithified.

Close to one million avalanches occur worldwide each year, and result in about 200 deaths. One in 10 victims of avalanches dies. Switzerland tops the avalanche fatalities in the world, and New Zealand and the Rocky Mountain U.S. and western Canada rank high in avalanche deaths as well.

First records of avalanches, in the Alps, go back about 2,000 years. The first known fatality, however, may be that of the Iceman whose brown, mummified corpse came to light in the Alps on the Austria-Italy border in September 1991. The corpse is about 5,200 years old. An avalanche may have buried the Iceman, the snow compacted into ice, and locked him in a glacier. Warm dry winds melted the ice cover to expose his body to an amazed world.

Snow avalanches, more common than ice avalanches, come in two main types: loose-snow and slab. Loose-snow avalanches are of low-density snow, like dry sand. Slab avalanches, the more dangerous, are of high-density snow and break loose in discrete slabs with a sharp fracture line.

Whatever the type, four ingredients come together to create an avalanche: steep slope, loose or slab snow, weak under-layer, and trigger. Most avalanches occur on slopes of 30 to 45 degrees, but can take place on slopes of less than 30 degrees and more that 60 degrees. A great depth of snow is not necessary, and bare ground may be exposed in places. But greatest danger exists just after a new snowfall. The weak under-layer is made up of sugar-snow crystals, caused by long-period freezing at the surface. When covered with new snow, the old sugar-snow crystals form an unstable under-layer on which snow may avalanche. A trigger for an avalanche may often be the weight of a skier or snowmobiler plus machine, and even a sonic boom or earthquake.

Personal Steps to Take Against Strictly Rock Landslides

1. Avoid living on, at the top, or at the base of a steep slope--especially one where cracks in rocks or rock layers, zones of weakness, are parallel to the slope.

2. If you are a homeowner on a slope or at its top, plant deep-rooted trees and shrubs to help stabilize the slope. Also, consider removing groundwater from a slope with drain tile or by pumping. Water adds weight to slope material and lubricates it.

3. Consider purchasing earthquake insurance if you own a dwelling adjacent to a steep slope in earthquake-prone places.

4. Avoid hiking on steep, rock debris-covered slopes, especially during or shortly after heavy rain.

5. Be aware of engineering measures taken to stabilize slopes. These include: 1) anchoring concrete slabs to solid rock at the base of slopes; 2) placing a metal mesh cover, like a chain-link fence, over a slope to restrain soil and rock debris and encourage plant growth; 3) burying concrete retaining walls within a slope; 4) removing soil and rock from the top of a slope to reduce weight there; and 5) reducing the slope angle.

6. Acquire more information on landslides by contacting the National Landslide Information Center at www.landslides.usgs.gov/index.html.

Personal Steps to Take Against Avalanches

1. Heed avalanche warnings and, if you have a choice, avoid

avalanche areas altogether. Trees pushed in the same direction often indicate an avalanche–prone slope.

2. Cross avalanche areas quickly, in the protection of trees and rock accumulations if possible. Travel with a group, and cross one person at a time.

3. Carry an avalanche location beacon, and make sure it is turned ON.

4. Carry also a probing pole and a shovel.

5. Watch for imminent signs of an avalanche: shooting cracks that stem from you, hollow sounds that emit from the snow as you ski or snowshoe, and rolling snow balls and pellets.

6. If caught in an avalanche, try to "swim" to stay on top. As you stop, place your hands and arms over your head to create an air space. Fight panic until you are rescued.

Additional Reading

Goodwin, Peter H. *Landslides, Slumps and Creeps*. Scholastic Library Publishing, 1998.

Chapter 16
Radon

Radon (RAY-dahn) is an odorless, tasteless, invisible, nonflammable gas that indirectly causes lung cancer. Estimates say that radon causes 15,000 to 22,000 deaths each year--more than drunk driving, falls in the home, fires, and drownings. Warnings from the Surgeon General point out that radon is the second leading cause of lung cancer in the U.S. Smoking is the leading cause. If you smoke and live in a dwelling with high radon levels, your risk of lung cancer skyrockets.

Radon forms by the radioactive decay or breakdown of the element radium. Radioactive decay is a natural process whereby one element decays or breaks down to form another.

The route from radon to lung cancer is rather devious. First we must step back to one type of the element uranium, the great-grandfather of radon, which decays to form the daughter elements thorium, protactinium, and radium. Radium was used at one time to make watch dials glow but was dropped for something less radioactive. Radium decays to form radon, the only gas in the chain of decay.

Radon, itself, is radioactive and decays to several heavy metals: polonium (two forms), bismuth, and lead (three forms). Of these daughter elements, polonium is notably troublesome. Fine particles of all these heavy metals stick to the lungs and to air passages leading to the lungs. As the metals decay, they release energy in three forms of radiation: alpha particles, beta particles, and gamma radiation, which is like X rays. Particularly insidious are alpha particles which strike lung cells and may damage them so that they become cancerous.

An ironic fact is that radon is relatively harmless by itself--you can breathe it in and out without detriment. But its sticky dangerous daughters, with their bursts of radiation, can do you in.

Geology of Radon

Radon occurs in outdoor air, indoor air, rocks, soils, and groundwater. Its ultimate source, uranium, is present in all rocks--in varying amounts--including the soils derived from the rocks. Some rocks contain more than average amounts of uranium: light-colored volcanic rocks, granite, dark shale, sedimentary rocks with phosphate, and metamorphic rocks formed from the above types of rocks. *Granite* (Figure 16-1) is a coarse-grained igneous rock rich in feldspar

Figure 16-1. Granite. Weathering takes place more so along fractures to produce blocks and rounded forms. Blocks are about 5 to 10 feet (1.5 to 3 meters) high.

and quartz (Figure 16-2) with minor dark minerals including flaky mica. *Igneous rocks* form by the cooling and solidifying of molten rock material. Rocks subjected to heat, pressure, and chemical action transform into *metamorphic rocks*.

You can gain a preliminary idea of how radon-prone your area is from a quick glance at a geologic map that depicts rock types.

Radon moves readily through fractures in rocks and through pore spaces in rocks and soils. Movement is faster where pore spaces are well-connected, as in sand and gravel, than where they are not, as in clay. Groundwater tends to slow down radon movement. From

Figure 16-2. Quartz and feldspar. The quartz fragment is 2.4 inches (61 millimeters) across.

these generalizations, you can expect homes with higher levels of indoor radon if they rest on: fractured rocks, drier rocks and soils with well-connected pore spaces, hill slopes, and the bottoms of canyons.

Radon Potential Where You Live

You can get a general idea of radon potential in your area from a radon potential map, one of which is the U.S. Environmental Protection Agency's (EPA) Map of Radon Zones of the U.S. The map portrays three zones: Zone 1, counties with a high predicted average indoor radon level; Zone 2, with a moderate radon level; and Zone 3, with a low radon level. You can view this map online at www.epa.gov/iag/radon/zonemap as well as view a radon potential map for each state. High levels of radon (Zone 1) can be expected throughout much of the Appalachian Mountains and New England, the Midwest, the north-central states, and the Rocky Mountain region. Low radon levels may be expected in the Atlantic coastal region, the Gulf States, and northern California and western Oregon and Washington. The Sand Hill region of west-central Nebraska is an island of low radon within a sea of high predicted values. Moderate radon levels may be expected in those regions not already mentioned.

Use radon potential maps only for general planning. Always test the specific radon levels in your dwelling with a reliable testing device (see page 86).

Several maps contribute toward the creation of radon potential maps: geologic maps, soil maps, radioactivity maps, and maps of indoor radon measurements. I've already mentioned how geologic maps may help, keeping in mind which rocks contain higher than the average amounts of uranium.

Soil maps may show the uranium or radium content of soils as well as the moisture content and interconnectedness of pore spaces in the soils. Others may display the radon content in the pore spaces of soils.

Radioactivity maps show radioactive energy that radiates from the ground. They are based on measurements taken with instruments from aircraft flying at low altitudes. These maps give a quick estimate of radon potential but have their limitations. Remembering that water blocks radiation, swampy and marshy areas will display lower than actual values of radioactivity.

Testing for Radon

Radon testing devices are of two main types, those canisters or bags with charcoal that absorb radon from the indoor air and alpha track devices that are struck by alpha particles from the decay products of radon. Follow the instructions that come with these devices, but keep in mind to place a tester in the lowest lived-in level, like a basement, and keep the doors and windows of a dwelling closed at least 12 hours before the test and during the test. Make sure to use a tester that meets EPA requirements or is state-certified. For results, send the tester to a laboratory recommended for the tester.

The EPA recommends a two-step approach for testing. First do a short-term test, usually for two to seven days, with one of the charcoal testers. The EPA sets the critical radioactivity level at 4 picocuries (PEE-coe-cure-eez) per liter of air (pCi/L). (For comparison, outside air normally contains about 0.4 pCi/L.) *Pico-* means one trillionth of something and *curie* honors Marie Curie, a French physicist who pioneered research on radioactive elements. If your value is less than 4 pCi/L, your dwelling is relatively safe and your testing is done--but keep in mind that some risk to your health still exists. If above 4 pCi/L, do another short-term test right away. Should the average of the two tests remain above 4 pCi/L, consider fixing your home to reduce the radon, and do a long-term test with an alpha track device for up to one year.

Your test results may contradict what you might expect from knowledge of the geology. I reside 2.5 miles (4 kilometers), as the magpies fly, from the base of a mountain that includes granite and similar rocks. Knowing granite contains higher than average amounts of uranium would lead you to surmise a high level of radioactivity in my home. But no. The last short-term test yielded a value of 2.1 pCi/L, well below the critical value. My previous home was at Grand Forks, North Dakota in the Red River Valley of the North, resting on clay and silt of prehistoric glacial Lake Agassiz. Three radon tests I made there gave radioactivity values several times greater than the critical value of 4 pCi/L! Simultaneous tests in January 1988 in the kitchen and basement gave results of 19.7 and 26.4. A repeated test in the basement in April recorded a value of 34.2.

Granite and similar rocks in northern Minnesota now lie buried under rock debris left by glaciers. Did glacial lake clay and silt incorporate radioactive materials from the chemical decay of granite-like rocks once exposed, flushed by westward-flowing streams into the lake?

Besides its presence in air, radon also occurs in water but the risk is much less. Well water is more apt to contain radon than a public water supply. If you are concerned about possible radon in your water, public or well, contact a laboratory that can measure radiation in water.

Reducing Radon in Your Dwelling

Radon that moves through cracks in rocks and pore spaces in rocks and soils finds its way into dwellings via cracks in the foundation, at floor-wall joints, piping trenches, and drainage systems. The dwelling creates a vacuum which sucks radon inside from the soil in two ways. During colder weather with the windows closed, warm air rises to create a vacuum in the lower parts of a dwelling. Suction draws air from the soil transferred into the basement. Various air exhaust openings and devices draw air out of a dwelling. These include flues on furnaces and chimneys for stoves and fireplaces and exhaust fans in kitchens and bathrooms. So if a dwelling naturally draws air through it, so will it draw radon.

You can reduce radon in your dwelling yourself with a do-it-yourself manual or by hiring a contractor. Whoever does it, the general approach is the same--called "sub-slab depressurization." Let's briefly consider this for an existing dwelling with a concrete-lined basement. (Modifications of this approach are necessary for dwellings with a crawl space or if applied to a water drainage system.) You break through the concrete floor at one or more places, insert a plastic pipe, seal the pipe well at the floor opening, and vent the pipe outside, well above the roof. Attach a special fan to the pipe that has good suction pressure, greater than the suction created by the dwelling. The fan runs continuously to vacuum out radon at all entry points before it invades the dwelling. Check the effectiveness of the radon reduction system with a short-term radon tester right after it's installed and then from time to time to make sure radon levels remain low.

Radon Contacts

For recorded information on radon, to request a brochure on radon, or to order a short-term test kit, call the National Safety Council's Radon Hotline, 1-800-767-7236.   To speak with an information specialist, call 1-800-557-2366.   The website www.epa.gov/iaq/contacts provides a contact for a state radon office.

Additional Reading

Burkhart, James, F., D.L. Kladder, and S.R. Jelinek, eds., *Protecting Your Home From Radon: A Step by Step Manual for Radon Reduction*. Colorado Vintage Companies, 1995.

U.S. Environmental Protection Agency. *A Citizen's Guide to Radon: The Guide to Protect Yourself and Your Family From Radon (4th edition)*. U.S. Environmental Protection Agency Document 402-K02-006), 2002.   (View this document at www.epa.gov/iag/radon/pubs/citguide.)

Chapter 17
Asteroids, Meteorites, and Comets

Astronomic invaders from space include asteroids, meteorites, and comets. All can impact Earth. *Asteroids* are rocky or metallic bodies smaller than planets. Most asteroids cluster in a belt between the planets of Mars and Jupiter and orbit the sun. Some pass out of this belt and cross Earth's orbit. A favored theory is that asteroids are remnants of a potential planet that failed to form.

Meteorites are mostly pieces of asteroids that impact Earth. *Meteors*, also called shooting stars or falling stars and identified by streaks of light, are meteorites that haven't struck Earth. Most meteorites burn up before reaching Earth's surface. Some meteorites may have come from Earth's moon and Mars, as identified from their mineral and rock makeup. Asteroids that slammed into the moon and Mars hurled fragments toward Earth whose gravity captured them.

Comets or "dirty snowballs" are made of stony cores covered by a layer of ice and rock. They're considered to be materials leftover after the creation of the sun and planets. Comets display tails when they pass near the sun, flashed by solar heat and solar wind.

Impact Craters

Earth has a long history of extraterrestrial impact. Ages of craters range from nearly two billion years to a few thousand years. Earth is scarred by more than 160 impact craters, most of which are less than 200 million years old. Many are of multiple hits from an asteroid or comet breaking into pieces upon its entry. An Earth Impact Database (www.unb.ca/passc/ImpactDatabase) lists the craters along with location, age, and images.

Around 160 impact craters may seem like a small number given Earth's long history for impacts to occur. But keep in mind that most space invaders of the past have struck the ocean because it covers nearly three-fourths of Earth's surface and hides most submarine craters. Also be aware that many ancient craters on land must have been destroyed by normal erosion. Many more impact craters are evident on the moon, Mars, and Venus because erosion, as it takes place on Earth, doesn't demolish them.

Let's look at a sample of Earth's impact craters.

Meteor Crater (also known as Barringer Crater) (Figure 17-1), west of Winslow in northeastern Arizona, is the best preserved of its size because of its relative youth and being little eroded in a desert environment. It also carries the distinction of being the first recognized impact crater on Earth. The crater is about 0.75 mile (1.2 kilometers) across and 600 feet (183 meters) deep. From the air, Meteor Crater is oddly more squarish than circular, its shape controlled by the orientation of joints, a northwest-trending set and a northeast-trending set. This crater was first thought to be that of an extinct volcano, but meteorite fragments have been found in the surrounding plain and by drilling within the crater itself. Further evidence of impact includes a crater rim 100 to 200 feet (30 to 61 meters) above the surrounding plain, up-arched rock layers, and ejected rock blocks some as large as small houses. Meteor Crater formed 49,000 years ago by a meteorite estimated at about 130 feet (40 meters) across. The meteorite may have struck at an angle from the southeast (from the right in Figure 17-1).

Figure 17-1. Meteor Crater. West of Winslow, northeastern Arizona. Notice the squarish outline, controlled by intersecting joints in the region. View is to the northwest. A road is visible to the right of the crater. Photograph by D.J. Roddy, U.S. Geological Survey.

Another well-authenticated impact crater is the Wolfe Creek Crater in northern Western Australia south of Halls Creek. Within a desert like Meteor Crater, and less than 300,000 years old, this Australian crater is also well-preserved. Wolfe Creek Crater is 2,870 feet (875 meters) across and 140 feet (43 meters) deep. Several large pieces of meteorites, up to more than 300 pounds (136 kilograms) have been picked up near the crater.

The largest known meteorite crater associated with meteorite debris is the New Quebec Crater at the north end of the Ungava Peninsula, 55 miles (88 kilometers) northwest of Kangiqsujuaq, in Quebec, Canada. It measures 1,128 feet (344 meters) across and

1,325 feet (400 meters) in depth. A lake 820 feet (250 meters) deep occupies the crater. The crater, well preserved in tundra, formed 1.4 million years ago.

You might wonder, how impact craters are identified, especially the older ones. First, a volcanic origin for a crater can be eliminated if the suspected impact crater occurs in non-volcanic terrain. And, besides finding meteorites nearby and other clues mentioned for Meteor Crater, still other bits of evidence arise from the high pressures and temperatures of impact. *Shatter cones*, cracked nested cones in rocks, are strong evidence for shock waves created by meteorite impact. They are most common in finer grained rocks like limestone and *quartzite*, which is metamorphosed sandstone.

Shocked quartz grains also support an impact event. Under a microscope such grains show parallel cracks across crystal faces in intersecting sets. Such features result when shock waves wield shearing stresses on quartz grains.

High pressure impact can create new minerals such as stishovite (STISS-hoe-vight) and coesite, both high-pressure forms of quartz. Stishovite often forms when high pressure shocks quartz. This special mineral can also form at considerable depth in Earth, where high pressures exist, but reverts to quartz before erosion brings it to the surface. Any stishovite at Earth's surface must have been formed by meteorite impact.

High temperatures can fuse some of the rocks upon impact. Natural glass, from fused quartz sandstone, attests to these high temperatures upon impact. Tiny glass spheres also may form.

Near-Earth Objects (NEOs)

Near-Earth objects (NEOs) are asteroids, meteorites, and comets in the neighborhood of Earth and capable of crossing Earth's orbit. The National Aeronautics and Space Administration's (NASA) search program continues to look for NEOs daily. As of March 2003, about 2,300 NEOs were known, and more than 600 were estimated at 0.6 mi (1 kilometer) across or larger. This web site gives a running account of the numbers: www.neo.jpl.nasa.gov/neo.

The force of an asteroid slamming into Earth is hard to imagine. Scientists estimate that an asteroid one or two miles (1.6 or 3.2 kilometers) across would strike Earth at a hundred times the speed of a high-powered bullet and with an explosive force of more than 100,000 million tons of TNT, 2,000 times larger than the largest hydrogen bomb built.

Several close calls or near encounters with asteroids have taken place. Events include those which occurred in 1937, 1989, 1996, and

2002, among others.   The asteroids were a mile (1.6 kilometers) across or smaller.   Distances from Earth ranged from a half million miles (0.8 million kilometer) to 75,000 miles (120,000 kilometers).

A well-known explosive "near-hit" took place in 1908 near the Stony Tunguska (toon-GOO-skuh) River in southern Siberia. I say "near-hit" because no impact crater formed. Evergreen trees were flattened over an area of 830 square miles (2,150 square kilometers), an area considerably larger than that of New York City and Washington, D.C. combined. Investigators found stony extraterrestrial particles embedded in the trunks of the fallen trees. Many reindeer were killed but no human casualties because of the sparsely populated region. The shock wave threw people to the ground or knocked them unconscious 37 miles (60 kilometers) away. Witnesses heard the blast from more than 60 miles (96 kilometers).   Disagreement still exists as to whether the intruder from space was an asteroid or a comet that exploded in the atmosphere maybe 4 miles (6 kilometers) above Earth's surface. In any case, witnesses saw a brilliant fireball 110 miles (170 kilometers) from the explosion whose smoke and dust led to colorful sunrises and sunsets in Europe and Central Asia.

One set of estimates says the asteroid was only 160 to 200 feet (50 to 60 meters) across and traveled at a speed of 9 miles per second (15 kilometers per second). It may have had the explosive force of a Hiroshima bomb or greater.   A Tunguska-sized explosion is expectable each 100 years, and a larger one the size of the largest hydrogen bomb, about once each 1,000 years.

Effects of Impact By a Doomsday Asteroid or Meteorite

The following scenario gives you an idea of what could happen from impact of a sizeable asteroid or meteorite, let's say 6 miles (10 kilometers) across.

First sign of a devastating asteroid or meteorite, if you are close, is the flash of a brilliant meteor. This is followed by a thunderous shock wave. Dust, from rock pulverized upon impact, rises in a plume--like that of a mushroom cloud from the detonation of a nuclear bomb--several miles (kilometers) high. (A crater created by an underground nuclear explosion is very similar to an impact crater.)

A tsunami (Chapter 12) from the impact scourges part of the planet. One such tsunami, if triggered by a smaller asteroid or comet one mile (1.6 kilometer) across in the mid Atlantic Ocean, could wipe out much of eastern coastal North America and western Europe.

Wildfires spread from the intense heat. Smoke blackens the skies at noon along with the dust from impact. In time, soot covers everything.   The darkened skies drop surface temperatures, bring about an "impact winter" that lasts for months or years. This impact winter is similar to a "nuclear winter" triggered by an extensive nuclear explosion and by fires set off by that explosion. Photosynthesis on land is curbed, followed by crop destruction and ensuing starvation. Plant plankton in the sea die off and their downfall ripples through the food chain. Perpetual dark skies also reduce evaporation from the oceans which, in turn, reduces rainfall. Drought follows with insidious dust storms.

The shock wave compresses and heats the air so that nitrogen burns to form nitrogen oxide gases that are toxic to plants and animals. These gases also block out sunlight. Precipitation falling through the gases forms another threat to life--acid rain.

Impact strips away protective ozone in the atmosphere. This removal exposes survivors to lethal ultraviolet radiation.

Severe impact on land triggers volcanic eruptions and major earthquakes. Extensive eruptions trap greenhouse gases that cause global climate to revert and warm the planet to excess.

Defense Against Asteroid or Comet Impact

Defense against asteroids or comets impacting Earth is in the realm of "Star Wars" technology. One method is to launch an anti-satellite nuclear missile against an invader threatening Earth's space. The strategy is to deflect the asteroid or comet with the exploding missile, as far out as possible, especially beyond the moon's orbit. But the exploding missile must not break apart the invader, whose pieces might be more difficult to defend against. Something like defending against a single rifle bullet versus the buckshot from a shotgun.

A modified approach for defense is to detonate a hydrogen bomb-like missile near an asteroid or comet, again without shattering either. If the radiation from the explosion is great enough, the intruding body would vaporize.

Personal Steps to Take Against Asteroids and Comets

1. Heed warnings about possible near-Earth objects impacting Earth.

2. Assemble emergency food, water, medical supplies, and drugs as you would for a possible attack by terrorists. Include a battery-powered radio to monitor an impending impact or what to do after impact.

Additional Reading

Erickson, Jon. *Asteroids, Comets, and Meteorites: Cosmic Invaders of the Earth*. Facts On File, Inc., 2003.

Chapter 18
Nuclear Power and Waste

Nuclear power is a clean source of energy compared to that generated from so-called fossil fuels. But nuclear power, produced from uranium, creates hazardous radioactive waste that may take thousands of years to decay.

Siting of Nuclear Power Plants

Siting of a nuclear power plant always requires detailed geologic studies. The supporting rock material must be strong enough to support the plant's structure. Landslide damage or destruction of the structure must be ruled out. The earthquake history in the vicinity of a possible site must be well-known. Since most earthquakes are caused by fault movement, they give clues to a possibly active fault at or near the plant site. Even a small amount of fault movement could fracture a nuclear reactor.

A nuclear power plant at Eureka, California was shut down in 1976 because of the potential of earthquakes. In 1980 a strong earthquake took place near the plant. Although undamaged, the plant remained closed for good.

Geology of Uranium

The metal uranium occurs in the heavy, black mineral pitchblende, found in veins that have been forced during a molten state to form igneous rocks emplaced beneath the surface. With erosion, pitchblende reaches the surface and reacts with the atmosphere. Soluble in water, uranium leaches out and moves by surface and groundwater, to be later absorbed by clay minerals and organic matter. Yellow carnotite is a new mineral so formed, common in the Colorado Plateau and concentrated in petrified wood and fossil bones of old stream channel sandstones.

Because organic matter absorbs uranium, other uranium sources include marine phosphate-rich rocks in Idaho and Florida and black shales in the eastern U.S. Prospectors check for uranium-bearing rocks with Geiger counters.

Disposal of Radioactive Waste

In Chapter 16 I mentioned that radioactive elements release hazardous radiation as they decay and form new elements. Uranium, of course, is amongst this group. As nuclear reactors produce electricity, they create high-level solid radioactive waste--mostly spent fuel rods--and low-level liquid and solid waste, which includes all other hazardous materials. Because of the long decay rates involved, radioactive waste may have to be stored with safety for thousands of years.

The majority of states do not encourage a nuclear waste storage site. A few agree if they are adequately compensated. Hot debates rage on this issue. Even the trucking of radioactive waste through states raises eyebrows and wrinkles foreheads.

Among the rocks that have been analyzed for radioactive waste storage are salt, shale, volcanic ash, and such igneous rocks as granite. Salt in thick layers is most desirable because it is unlikely to crack and can absorb much heat, another product of radioactive decay.

Yucca Mountain, 110 miles (177 kilometers) northwest of Las Vegas, Nevada is the likely site for the disposal and storage of high-level waste. It would harbor more than 100 miles (161 kilometers) of tunnels. The rock at this site is lithified volcanic ash. Although chosen by the U.S. Congress in 1988, the site is still being analyzed. Desirable traits include its remote location, a water table 1,500 feet (457 meters) below the surface, and annual rainfall of only 3 to 5 inches (7.6 to 12.7 centimeters). But a fault lies under the site and another borders it. An earthquake struck 12 miles (19 kilometers) from the site in 1992. Yucca Mountain was created from volcanic eruptions 12 to 15 million years ago, and a volcano lies 11 miles (18 kilometers) away. Volcanic activity and faulting could expose radioactive waste to the atmosphere and raise the water table.

Groundwater contamination is a particular problem from radioactive waste. A seasonal rise in the water table must not cover the waste for thousands of years even though the disposal site is deep underground. At Yucca Mountain a water table rise from volcanism could contaminate the groundwater. Active faults could allow contaminated groundwater-heated by radioactive decay--to rise through them.

Besides the radiation hazard, heat pollution and terrorist activity must be dealt with from nuclear reactors. Two-thirds of the heat generated at nuclear power plants is wasted heat that must be disposed of. If not released with care into streams, lakes, and along seacoasts, heat may decimate aquatic creatures. Terrorist sabotage of a nuclear plant by explosive or other means may wreak long-lasting havoc on a large population.

Personal Steps to Take With Nuclear Power and Waste

Red flags went up with accidents at the Three Mile Island

reactor in Pennsylvania in 1979 and the Chernobyl reactor near Kiev in the Ukraine in 1986. As of this writing, 104 licensed nuclear power plants in the U.S. continue to provide electricity, and a nuclear accident could happen again. Storage of spent fuel rods is now temporary, in water pools at the plants, and the Yucca Mountain site for permanent storage has not yet been approved.

Steps you can take with nuclear power and waste include:

1. If you live near a nuclear power plant, consider moving or work out a plan for your escape in case of an accident or a terrorist event. The International Nuclear Safety Center provides a map of North American nuclear power reactors at this web site: www.insc.anl.gov/pwrmaps/map/north_america.php. Most are in the eastern U.S.

2. Consider opposing the construction of new power plants, at least until a high-level waste site is open to accept waste.

3. Provide your input into the approval of a permanent radioactive waste site based on your knowledge of what geological conditions are most desirable.

Chapter 19
Rise in Sea Level

Rises in sea level can take place short term or long term. Short-term rises occur in a matter of feet per hour or day, long-term rises on the order of inches per century. In either case, noteworthy destruction is possible.

Short-term Rise

Hurricanes bring about short-term rises in sea level called *storm surges*. Hurricanes are strong winds, to more than 200 miles (322 kilometers) per hour, that rotate counterclockwise, in the northern hemisphere, around a region of low air pressure. This low pressure in the hurricane's center sucks up the ocean's surface into a dome or local sea-level rise which is pushed shoreward by the hurricane. Strong onshore winds, and especially if timed with a high tide, emphasize the pile-up of water onshore. Storm surges, common along the U.S. Atlantic and Gulf Coasts, can readily raise sea level 15 feet (4.6 meters) or more. This is significant in light of the fact that much of the populated Atlantic and Gulf coastline is less than 10 feet (3 meters) above sea level.

More people are killed by storm surges than by hurricane winds. Coasts are especially vulnerable where ocean floors are shallow and gradual. The worst storm surge this century was 24 feet (7.3 meters) high, pushed by Hurricane Camille into Pass Christian, Mississippi in 1969. More than 18,000 homes and 700 businesses were destroyed or seriously damaged. Camille, rated at the highest category 5, and the storm surge together killed 256 people in Mississippi and Virginia.

A storm surge 8 to 15 feet (2.4 to 4.6 meters) hit Galveston, Texas in 1900 and inundated entire Galveston Island. Combined storm waves and the storm surge wrecked numerous buildings and killed 6,000 people, many of whom drowned. Estimated damage was $30 million.

Cautionary steps have been taken for storm surge warnings. A MEOW (Maximum Envelope of Water) warning is a computer warning for essentially all places along the U.S. Atlantic and Gulf Coasts. This warning is based on the likely volume of water pushed onshore by a hurricane. A SLOSH (Sea, Lake, and Overland Surge) computer model also evaluates a given storm surge threat. A web site of the National Hurricane Center, www.nhc.noaa.gov/HAW2/english/storm_surge , provides safety and other information about storm surges.

Long-Term Rise

A little background might be desirable before we explore long-term sea-level rise. Within the past 2 to 3 million years, sea level, world-wide, has risen and fallen several times. This rise and fall corresponds to global warm and cool periods, which relate to the melting of glaciers on the one hand and their growth on the other. So sea level rises when glaciers melt and falls when glaciers build up and spread over the land. About 18,000 years ago, when glaciers last reached their greatest extent, the sites of such northern cities as Chicago, Detroit, and Montreal lay beneath thousands of feet of glacier ice. Since that time glaciers have generally been retreating and most glaciers are melting today.

In the past 15,000 years, sea level has risen about 400 feet (122 meters). The rate of rise began at 4 feet (1.2 meters) per century, then slowed to 1.5 inches (3.8 centimeters) per century. Of late, some say sea level has been rising 10 to 12 inches (25 to 30 centimeters) per century along much of the U.S. coast, others say 4 to 10 inches (10 to 25 centimeters). Whichever numbers are closer to the truth, the rises are significant and the reasons for the rises are uncertain. In Chapter 7, I mentioned that Antarctica holds 90 percent of the glacier ice, and Greenland contains essentially all of the rest. Should the existing glaciers all melt, sea level would rise more than 300 feet (100 meters). With such a rise, the Statue of Liberty's torch in New York Bay would barely be above water, if at all!

You can view maps of the U.S. Atlantic and Gulf coasts vulnerable to sea level rise at this web site: www.yosemite.epa.gov/OAR/globalwarming.nsf/content/ResourceC enterPublicationsSLRMaps. Regions are color-coded as follows: red, those affected by a sea-level rise below 5 feet (1.5 meters); blue, by a 5 to 11.5 feet (1.5 to 3.5 meters) rise; and yellow, by a rise above 11.5 feet (3.5 meters). Louisiana, Florida, Texas, and North Carolina account for more than 80 percent of the low land vulnerable to coastal flooding. Sea-level rise on low-lying coasts allows flooding well inland. A small rise may translate to ocean water encroaching several miles landward. A rise of just a few feet could demolish thousands of oceanfront houses and hotels.

On steep coasts with sea cliffs a sea-level rise quickens erosion. Cliffs are worn away by the battering of ocean waves and by landsliding.

Debate continues as to the cause of warming that is melting the glaciers. Some scientists say it's part of a cyclical pattern that has existed for millions of years. Others believe that present warming is due to the "greenhouse effect." The increased burning of coal and oil releases surplus carbon dioxide and other gases that trap the sun's energy and warm the air and oceans. As ocean water warms, it expands and contributes to the problem of sea-level rise.

Personal Steps to Take Against Sea-level Rise

For short-term sea-level rise along coasts:

1. Plan your best route to reach safety.

2. Arrange to stay at the dwelling of a friend or relative, or seek a motel or public shelter.

3. If your dwelling is in jeopardy, board up the windows, turn off the utilities, including water, and lock it.

4. Fill your vehicle with gas and withdraw sufficient cash.

5. Take any prescription drugs with you.

6. Stay tuned to local weather with a battery-operated radio.

For long-term sea-level rise:

1. Consider re-locating if you have oceanfront property.

2. Monitor the debate about global warming and the present trend of glaciers to melt.

Additional Reading

Titus, James G., and Vijay Narayanan. *The Probability of Sea Level Rise*. Diane Publishing Company, 1998.

PART 3. GEOLOGY TO STIR THE BRAIN

Chapter 20
Two Unifying Concepts

Two unifying concepts in geology tower above others: the immensity of time and plate tectonics. Many believe that the concept of the immensity of time has been geology's greatest contribution to general thought. *Plate tectonics*, the concept that Earth's outer rind consists of rigid rock plates that constantly shift and interact with one another, explains the distribution of major earthquakes, volcanoes, and mountain belts.

Immensity of Geologic Time

To we humans, time is something we say we both "make" and "kill." Time ticks away without our making it, and we certainly don't want to kill it. Finite for us, time is something we should cherish. Without time, we accomplish nothing. When young, we revel in presumed immortality. With age, we cast a watchful eye on the remainder of our moments on Earth.

To Earth, time is open-ended, limitless. We measure human life in tens of years, but Earth's existence in terms of *billions* of years. Historic time, that period of written history, goes back only about 5,000 years. Prehistoric time precedes historic time and that since Earth's origin can be called *geologic time*. Bear in mind that the universe, of which Earth is a small part, had to have originated billions of years earlier.

Geologic Time Scale. Subdivision names of the geologic time scale (Table 20-1) come from a variety of sources including places, rocks, a mountain range, and ancient tribes. The Silurian, for example, stems from Silures, an ancient Welsh tribe.

Geologic time is first divided into eons, the Early and Late Precambrian (pree-CAM-bree-uhn) and the Phanerozoic (fan-uh-ruh-ZOE-ick). Eons are most uneven in duration for the Precambrian makes up more than 85 percent of geologic time. The Phanerozoic Eon is divided into the Paleozoic (pay-lee-uh-ZOE-ick), Mesozoic (mehz-uh-ZOE-ick), and Cenozoic (sehn-uh-ZOE-ick) Eras; eras into periods (such as Tertiary); and most often the Cenozoic into epochs such as Paleocene (PAY-lee-uh-seen). Notice that we live in the

Holocene Epoch. Boundaries of time terms tend to correspond to such events as the appearance or extinction of organisms and mountain building.

Table 20-1. Geologic Time Scale

Eon	Era	Period	Epoch	M Yrs Ago
Phanerozoic	Cenozoic	Quaternary	Holocene	0.008
			Pleistocene	
				2
		Tertiary	Pliocene	5
			Miocene	
				24
			Oligocene	
				34
			Eocene	
				54
			Paleocene	
				65
	Mesozoic	Cretaceous		
				145
		Jurassic		
				213
		Triassic		
				248
	Paleozoic	Permian		
				286
		Pennsylvanian		
				325
		Mississippian		
				360
		Devonian		
				410
		Silurian		
				440
		Ordovician		
				505
		Cambrian		
				544
Late		Precambrian		
				2500
Early				

Numbers at time term boundaries are after the U.S. Geological Survey, 2003. The time scale begins at 4,600 million years ago.

The geologic time scale begins at 4.6 billion years. Why? Although the oldest known rocks, from western Australia, are younger--more than 4 billion years old--the oldest moon rocks and meteorites have been dated at 4.6 billion years. A favored theory mentioned in Chapter 17 is that asteroids (and meteorites), clustered in an orbiting belt between Mars and Jupiter, are parts of a planet that failed to form. A unity exists within the solar system, and the history

Table 20-2.   Selected Life Events During the Phanerozoic Eon

Period	Epoch	Selected Life Event
Quaternary	Holocene	Last Passenger Pigeon, Red Colobus
(two epochs)		Monkey
	Pleistocene	First Modern Humans, Last Mammoths
	Pliocene	First Humans
Tertiary	Miocene	First Abundant Grasses
(five epochs)	Oligocene	First Monkeys, Largest Land Mammals
	Eocene	First Rhinoceroses, Mammoths
	Paleocene	First Horses
Cretaceous		Last Dinosaurs, First Flowering Plants
Jurassic		First Birds
Triassic		First Dinosaurs, Mammals
Permian		Last Trilobites
Pennsylvanian		First Cone-bearing Plants
Mississippian		First Reptiles
Devonian		First Amphibians, Insects
Silurian		First Confirmed Land Plants
Ordovician		First Land Animals (Millipedes?)
Cambrian		First Fishes, Trilobites, Corals

of its parts are entwined. So the age of Earth is the same as the age of the oldest moon rocks and meteorites.

Various natural clocks have been used to set up the geologic time scale. Earth, itself, behaves like a gigantic clock in its complete rotation every 24 hours. Fossils are organic clocks based on their steadfast progression through time. Table 20-2 lists notable biological events with time intervals. Rocks, of course, also clock time. The principle of *superposition* enters here: That in an undisturbed vertical sequence of rock layers, the oldest lie at the bottom, the youngest at the top. You can also reckon oldest versus youngest fossils by their position in the rock layers. Rocks, too, contain radioactive clocks, which permit geologists to measure time in years rather than just stipulate oldest versus youngest; that is, reckon time in an absolute rather than in a relative sense.

Measurement of Geologic Time. In Chapter 16, I mentioned that radioactive decay is a natural process whereby a radioactive element breaks down to form another. The rate of decay for all varieties of radioactive elements is known, seems to be constant with time, and presumably is not altered by changes in temperature, pressure, and chemical affects.

To date a rock, sophisticated equipment measures the amount of an original (parent) radioactive element and the amount of a derived (daughter) element. Applying the rate of decay gives the age of the rock. Suppose that with a common dating method, in which uranium 235 (the number refers to the atomic weight) decays to lead 207, 50 percent of the uranium has decayed to 50 percent lead. The decay rate of uranium 235, the time for half of it to decay, is 713 million years or the age of the sample. If the sample contained 25 percent uranium and 75 percent lead, the age would be 1,426 million years, the time for two successive halves of the uranium to decay.

Other common radioactive methods used to date minerals and rocks are based on the decay of rubidium 87 to strontium 87, potassium 40 to argon 40, and carbon 14 to nitrogen 14. The first three are used to date mostly igneous and metamorphic rocks, some of the oldest because of the long decay rates involved. On the other hand, the carbon 14 method can be applied only to materials that contain carbon such as wood, bone, and shell. And because of the rapid decay rate, this method can only date materials younger than about 70,000 years.

You might ask, are the ages based on radioactive decay trustworthy? Having been tested by generations of scientists, they stand firm. If a calculated age is suspect, doesn't fit in with others or contradicts the principle of superposition or what fossils indicate, the test is re-run or a new sample is analyzed. Sometimes contaminated samples give false dates. Any new date must agree with a scheme established by other dates as well as by the rock and fossil records.

Back to the geologic time scale. It developed during the eighteenth and nineteenth centuries mainly in Europe, the birthplace of geology, by first stacking up rock layers in one place and matching them with rock sequences in other places. But rock layers change from place to place. So geologists turned to fossils to date rocks in a relative sense and worked up a time scale that could be applied over wide distances. With the coming of *radioactive dating*, ages in years were applied and the geologic time scale was calibrated. This calibration continues.

Comprehending the Immensity of Geologic Time. We gain some sense of the immensity of geologic time by realizing the duration necessary for certain events: the cutting of deep canyons such as the Grand Canyon; the multiple advance and retreat of glaciers; the creation of mountain ranges; and the flooding of seas on the continents and their subsequent withdrawal, which has happened many times in the geologic past. Such events require millions of years as inferred by the principle of *uniformitarianism:* That Earth processes operating today have also operated in the past because natural physical and chemical laws have not changed through time. The pull of gravity caused streams and landsliding to cut canyons, and glaciers to cut glacier valleys; side-to-side squeezing has caused mountain ranges to rise; and rise and fall of sea level has caused the flooding and withdrawal of seas from the continents.

To comprehend geologic time on a human level, consider that many persons today live to an age of 100. If because of a medical breakthrough you lived ten times as long, you would be a 1,000-year-old Methuselah. Ten times longer would be 10,000 years old. A real geriatric but still a youngster as compared to the duration of many geologic events. If your imaginary age were increased ten times, two or five times more, you would have reached a million or billion years. Even at a billion years, your age would pale in terms of Earth's four and a half billion years.

In Chapter 21, I ask if dinosaurs and humans were contemporaries. Fossils show that dinosaurs appeared on Earth about 225 million years ago, and reigned for 160 million years before becoming extinct 60 million years before we humans showed up. The long history of dinosaurs may seem striking, but not so in terms of the oldest known, fish-like vertebrates that swam 530 million years ago.

Plate Tectonics

In the concept of plate tectonics, rigid Earth plates pull away from each other, jam together, or slide alongside one another. You might compare these plates to ice slabs that slide, slam, and slip under one another during ice breakup in a stream in spring. The word "tectonics" refers to large-scale deformational features such as mountain belts.

Earth's Interior. Before getting into the details of plate tectonics, let's glimpse at Earth's interior and how thick the plates are in comparison. Deep-hole drillers have barely pricked the outer part of Earth to a depth of about 8 miles. So what geologists know of its interior is based on the behavior of seismic waves (Chapter 11) set off by earthquakes or human-created explosions. Seismic waves help reveal Earth's interior this way: They travel faster through denser rock material, and certain waves do not pass through liquids.

A boiled egg, although oval rather than nearly spherical, is zoned like Earth. The eggshell corresponds to Earth's *crust*, which averages 4 miles (7 kilometers) thick in the ocean basins and 25 miles (40 kilometers) thick under the continents, and is thickest at mountain belts. Basalt largely makes up the denser oceanic crust and granite probably constitutes most of the less dense continental crust. The egg white relates to Earth's *mantle*, down to a depth of 1,800 miles (2,900 kilometers). Seismic waves show that the mantle is

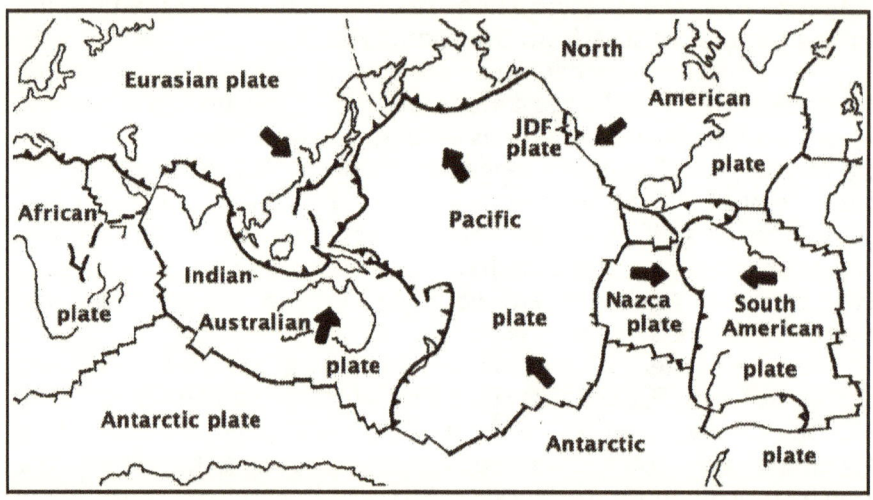

Figure 20-1. Major Earth plates of the plate tectonics concept. JDF off the northwestern U.S. coast represents the small Juan De Fuca plate. Arrows indicate the direction of movement of the plates.

denser than the crust and may be rich in iron and magnesium minerals. The egg yolk is reminiscent of Earth's *core*, with a diameter of about 4,300 miles (7,000 kilometers). Seismic waves reveal that the core is very dense like that of iron with maybe some nickel. The inner core seems to be solid but the outer six-tenths is likely liquid because certain seismic waves do not pass through that part.

Geologists believe that the plates of the plate tectonics concept--eight large plates and a few dozen smaller ones (Figure 20-1)--are made up of the crust and the uppermost mantle. They average about 60 miles (100 kilometers) thick and are thickest at the continents. Most plates include both continental and oceanic crust, such as the North American plate, but may consist only of oceanic crust, such as for the Pacific Plate. Plates slide on a hotter, weaker part of the mantle that extends to a depth of 125 miles (200 kilometers), and may be partly molten.

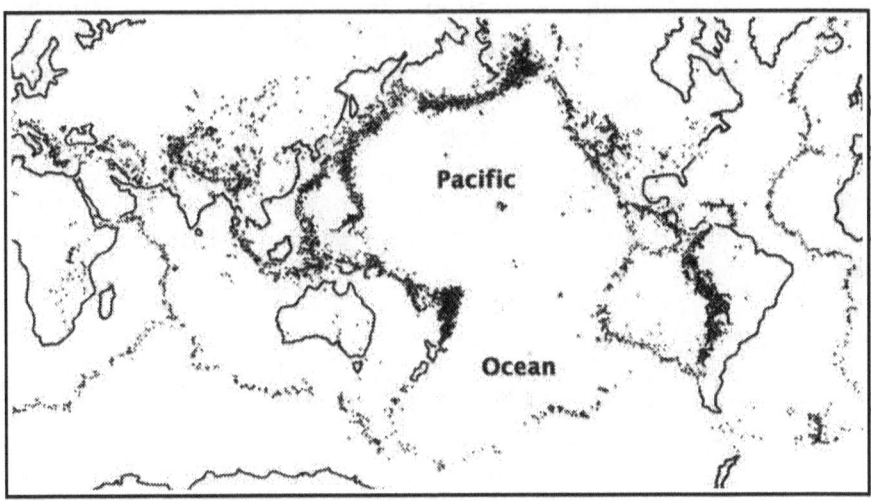

Pacific

Ocean

Figure 20-2. Distribution of earthquakes. Compare the occurrence of earthquakes to the boundaries of the plates in Figure 20-1.

Evidence for Plates and Their Movement. A precursor of the plate tectonics concept was the theory of *continental drift.* That less dense continents have drifted through more dense ocean basin rock, a different process than plates of both continent and oceanic crust and upper mantle sliding over a mushy mantle zone . The earlier fit of continents now separated supported this theory. How well, for example, does the eastern bulge of South America conform with the western indentation of Africa.

Similar rocks and fossils in now-separated southern continents also provided strong evidence for continental drift. Belts of folded rocks at the southern tip of Africa trend east-west and end at the southwestern coast. Similar folded rocks of the same age can be identified near Buenos Aires, Argentina. A peculiar fernlike land plant, *Glossopteris* (glahs-AHP-tuh-ruhs), has been found in rocks of the same age from Antarctica, Australia, India, South Africa, and South America. These occurrences strongly support the contention that these continents were once connected.

Another line of evidence that supports continental drift but more directly plate tectonics is seafloor spreading away from oceanic ridges. Iceland straddles the mid–Atlantic ridge and the amount of spreading can be measured directly across rift valleys (Chapter 9). The amount of spreading translates to the speed of plate movement, less than an inch (about 1 centimeter) to many inches (many centimeters) per year, depending on the location. Deep–sea sediments gradually thicken and become older away from both sides of oceanic ridges, which also testifies to seafloor movement.

The global distribution of earthquakes (Figure 20-2) is not haphazard but follows distinct belts. Major volcanoes occur in close proximity with the belts of earthquakes. Because of the dynamic interactions of plates where they meet, most geologists believe that most earthquakes and major volcanoes outline the boundaries of plates.

Where Plates Move Away From One Another. At oceanic ridges, pulling–apart stresses produce faults and rift valleys. Faulting causes earthquakes that align in belts along the trend of the ridges. Basalt lava issues along the faults and creates new oceanic crust. On continents rifting may eventually form ocean basins. We might perceive an east African rift valley, the Red Sea, and the South Atlantic Ocean as three stages in the formation of an ocean basin.

Where Plates Collide With One Another. Imagine a plate with continental crust colliding with one bearing oceanic crust (Figure 20-3). Thousands of feet of sediment accumulate along the juncture of the two plates. The heavier oceanic plate descends beneath the lighter continental plate. A good example today is where the oceanic Nazca plate descends beneath the continental portion of the South American plate at the Peru–Chile Trench. The colliding plates squeeze the sediments into a huge linear welt, which is faulted near the surface where the lithified sediments are brittle and bent or folded at depth where they are subjected to heat and pressure.

The huge welt of deformed sediments becomes a mountain belt like the Andes of western South America. Beneath the folded rocks, where heat and pressure are higher still, metamorphic rocks form. When the descending plate reaches depths where temperatures are very high, part of it melts. Some of the molten rock reaches the surface to form volcanoes, and some of it forces its way into all of the overlying rocks to solidify into huge bodies of rock like granite beneath the surface.

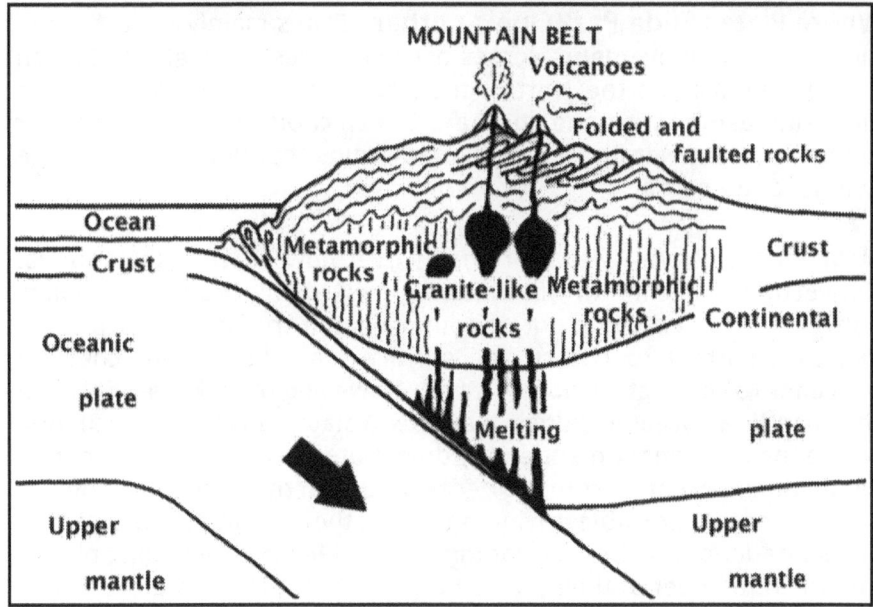

Figure 20-3. Collision of an oceanic and continental plate, and features associated with a mountain belt. This situation is similar to the collision of the Nazca and South American plates with the formation of the Andes Mountains.

Active mountain belts, like the Andes, coincide with zones of earthquakes and volcanoes. Older mountain belts are gradually worn down. Eventually, with the faulted and folded rocks eroded away, only mountain roots remain--of igneous and metamorphic rocks. Such extensive rocks in Voyageurs National Park, northern Minnesota, for example, testify to the roots of ancient mountain ranges in a place now devoid of mountains.

Where both opposing plates contain oceanic crust, one plate descends beneath the other, and metamorphic and granitic rocks may or may not form. The Pacific plate in collision with part of the Eurasian plate is such a place with an oceanic trench (Japan) and volcanic islands.

Where two plates bear continental crust, neither plate descends or very little because of the buoyancy of the crustal rocks. The Himalayas occur where rocks are bent and broken from the collision of plates both of which bear continental crust. They formed about 100 million years ago when India jammed part way under the Eurasian plate.

Where Plates Slide Past One Another. Plates mainly shift in side-slip fashion in movements across oceanic ridges. And along the San Andreas fault zone the North American and Pacific plates move in side-slip fashion with one another. Conspicuous earthquakes occur along side-slip margins as for plate margins that pull apart or come against one another.

How Plates Move. Debate remains over how or why plates move. Convection currents in the mantle may drive them. Visualize convection currents in a pot of soup. Heat at the bottom of the pot causes a current to rise to the surface. As the current cools, it descends to complete a loop of travel. Upwelling of rock material from the mantle at oceanic ridges may push plates apart, or circulation within the uppermost mantle may drag plates along. Heat for mantle circulation may come from the decay of radioactive minerals at depth.

Another possible driving process is the combination of slab pull and slab suction. Slabs, the leading edges of heavy descending plates, pull the remainder of their plates behind them. Or, the slabs create a downward suction force as they sink, like that when you pull a plug from a filled bathtub.

A Caveat

The evidence for plate tectonics is most convincing, and it has caused a revolution in geology much as the theory of evolution has revolutionized biology. Most geologists believe that plates exist and move to explain the occurrence of major earthquakes, volcanoes, and mountain belts. But, as has been said, science should not be swayed by the majority. We must also listen to the skeptical minority to further science. The theory of plate tectonics may be incorrect, at least in part. An example: How to explain the numerous earthquakes at the interiors, not at the margins, of plates?

Additional Reading
Erickson, Jon. *Plate Tectonics*. Facts on File, Incorporated, 2001.

Chapter 21
Asking Dumb Questions
About Geology–and Their Answers

Let's first clarify something: Dumb questions about geology don't exist. Some questions may seem "dumb" at first, but all questions lead to other questions that are thought provoking and revealing. As we've been told since we were children, "You learn by asking questions."

In this chapter I've assembled 20 questions that I believe you might ask, along with my answers. They are arranged within two groups, Earth Workings and Earth History. I hope that these questions will provide the impetus for you to ask others. Pick a geologist's brain. Most don't mind.

Earth Workings
Q 1: Why are the oceans salty?
A: When rocks interact at Earth's surface with air and water, they decompose into new minerals and various kinds of salts that dissolve in water. These salts are flushed away by rainwater and carried by streams, eventually to the oceans. Decomposition of rocks, also called chemical weathering, has been going on for eons. Over time, streams have built up the salts in oceans to make them salty. Sodium chloride or common table salt, the most common salt in sea water, forms by sodium from rock weathering combining with chlorine from volcanic gas. Other salts in sea water, from the decomposition of rocks, form by combination with calcium and magnesium.

Q 2: What formed the mile–deep Grand Canyon?
A: The Grand Canyon in northwestern Arizona lies at the western end of the Colorado Plateau, a crucial point to the canyon's origin. To find the answer to its origin, look to the Colorado River at the canyon's bottom. Indeed, the river cut the canyon. Downcutting began 8 to 10 million years ago, concurrent with the uplift of the Colorado Plateau, with much of the incising having taken place within the last 2 to 3 million years. Cutting of the mile–deep (1.6–kilometer–deep) canyon would have been impossible without the incessant uplift of the Colorado Plateau. Uplift kept the Colorado River's channel slope steep enough for active downcutting.

Some may stand at either rim of the canyon and speculate that the immense chasm yawned open abruptly along a major rift like a joint or fault (Chapter 9). But evidence for such a break doesn't exist. Joints or faults are almost always straight or slightly curved. A peek at a map of the Grand Canyon shows a canyon with many curves and bends--what you might expect if the canyon were created by the river.

Q 3: How can some streams cut through mountains instead of going around them?
A: The most common explanation is that a stream first flows over a mountain range that is buried in sediment. As the stream erodes into the sediment cover, it is let down or superimposed on the once-buried mountain range, and eventually cuts a canyon through it. Uplift of a region usually steepens a stream's channel and leads to more downcutting.

A good example involves the Wind River and the west-trending Owl Creek Mountains, a kind of western tail of the Bighorn Mountains in north-central Wyoming. While the mountains were once buried in sediment, the early Wind River flowed above them on gently sloping terrain. About 2 million years ago the climate became wetter, and was accompanied by uplift. The Wind River downcut its channel to eventually incise 3,000-foot-deep (900-kilometer-deep) Wind River Canyon as it kept up with the uplift. A drive on U.S. Highway 20 today is geologically rewarding. You witness an array of older to younger rocks as you proceed northward, with all of the main rock formations labeled.

A less common explanation of why streams may cut through mountain ranges is that a stream is older than a mountain range, its course established before a range exists. As a mountain range forms, maybe by forces squeezing against one another, a major stream may be able to maintain its downcutting through the new and rising mountain range.

Q 4: How does granite form? Why is it good for tombstones?
A: Granite, and similar rocks, makes up most of the outer rind or crust of the continents. Mineral crystals, mostly of feldspar and quartz but also dark minerals like flaky mica, in the rock are large enough to easily see without a hand lens. Depth is necessary for the slow cooling of molten rock, slow enough for large mineral crystals to grow. We see granite at the surface because erosion has stripped away the overlying rocks.

Granite is a good choice for tombstones because it generally weathers slowly. But it is not invincible, especially in humid climates. Tombstones of limestone or *marble*, heat- and pressure-altered limestone and dolostone, however, weather even faster in humid climates. Given sufficient time, on the order of millions of years, even a granite tombstone reduces to a smear of rusty clay. But none of us has to worry about this eventuality.

Q 5: In temperate regions, what keeps streams flowing after the snow melts?
A: As snow heats up, some of it passes directly into water vapor. Much of that which doesn't, converts to meltwater and finds its way into streams, ponds, and lakes. Some of the meltwater, along with rainwater, sinks into the ground to become groundwater (Chapter 6), which fills pore space in sediment and cracks in rocks. Groundwater moves from areas of high water pressure to areas of low pressure, such as along streams. Groundwater may feed streams long after snow melts, even year around.

Q 6: Will we ever run out of oil and gas?
A: Yes, of course. Oil and gas, as for Earth's other natural resources, are finite, nonrenewable. Once you use them up, they're gone forever, and can't be replaced. The world's resources of oil, that discovered and undiscovered, may last well over a 100 years or more. The U.S., however, is running out of oil at a fast pace, and imports nearly half that it requires. Either this country will need to seek out other fuels or sources of energy or import more oil.

A problem exists with the recovery of oil from the ground. The oil may be in too small of a concentration or too far from the market to make economic sense. Some may be under public lands, like national parks, and is off limits to drilling. And, in spite of good extraction techniques, maybe only 30 to 40 percent can be pumped to the surface. New technology, like injecting oil wells with carbon dioxide, may jack up the amount of oil that is recoverable.

To better perceive the reality of finite oil and gas resources, let's glimpse into how they form and accumulate. Microscopic animals and plants settle to the bottom of a lake or sea and build up organic matter in mud for maybe thousands of years. Other sediments bury the organic-rich mud, and all convert to rock. Heat and pressure from deep burial transforms the organic matter in the shale (once a mud) to oil and gas.

Pressure squeezes oil and gas out of the shale into a porous reservoir rock like sandstone and limestone that allows oil and gas to move through it. Oil and gas are more buoyant than water, and tend to rise through water-saturated reservoir rocks until they are trapped or contained and can no longer move. One way oil is trapped is at the crest of a an anticline beneath the surface. Geologists seek out the trapped oil and gas in reservoir rocks.

A couple other oil resources are possible if extraction is profitable. Oil or tar sands contain asphalt, allowing them to be mined as well as drilled. Rocks must be heated to release the heavy oil or tar. The Athabasca region of northern Alberta contains a notable oil and tar sand resource.   Oil shale also contains solid organic matter. Colorado, Wyoming, and Utah store significant deposits of oil shale which was formed from organisms in ancient lakes. Distilling the oil from the shale is not economical at present. Disposing of the spent shale, which expands during distillation, is a problem.   And the process requires much water, a premium product in arid places. Burning the shale in underground caverns might offer an alternate method of separating the oil from the rock.

Q 7: Why are the volcanoes in the Cascade Range lined up, north to south?
A: I'm glad you brought this up. I mentioned this fact in Chapter 13 but didn't say why.   To review, 15 major composite volcanoes extend north-south from southern British Columbia to northern California. Mt. Ranier, Mt. St. Helens, and Lassen Peak are among them.

We call upon the plate tectonics concept to explain the alignment of the volcanoes. In the vicinity of the Cascades, a small plate, the Juan de Fuca plate (Figure 20-1), meets and descends beneath the North American plate--according to the concept. At a certain depth, part of the Juan de Fuca plate melts. The rising molten rock is released at the surface as lava and volcanic rock fragments to produce composite volcanoes, lined up along the descending plate. The downward-moving plate may also rupture to set off earthquakes, such as the Nisqually earthquake epicentered between Seattle and Olympia, Washington in 2001.

Q 8: What are the pros and cons of building dams on streams?
A: Pros include generation of hydroelectric power, flood control, water storage, and recreation, which includes fishing, boating, and other water-related activity. Only the larger dams can generate hydroelectric power. Flood control is often the primary benefit. Water storage is significant in arid regions where selective release is necessary for irrigation and high-demand urban use. Recreation is often a tag-on benefit.

Cons include cost, the permanent flooding of fertile land, loss of recreation, and certain geological effects. Creation of a huge reservoir may destroy tens of thousands of acres of fertile agricultural land once on the original floodplain. Touted recreational use after a dam's emplacement might imply no recreation was available before. But fishing, hunting, and nature study can all be enjoyed along a free-flowing stream.

From a geological viewpoint, dams cause detrimental effects. Reservoirs, in due time, fill with sediment and die. They are temporary features of the landscape, just as are natural lakes. Life expectancy of most large reservoirs is on the order of 100 to 200 years. One example of a dead, useless reservoir is Mono Reservoir in Los Angeles, California, filled to the brim with sediment. Because reservoirs catch sediment from a dammed stream, water that issues downstream is clear. The clear water tends to erode faster below the dam and may cut into stream banks and attack bridges.

Q 9: I've heard that Los Angeles and San Francisco are moving closer together. This sounds like a tall story.

A: Though it may sound tall, I'm afraid it's true. But don't worry. Extra congestion from the merging of the two cities won't happen soon. Los Angeles and San Francisco rest on opposite sides of the San Andreas Fault (Chapter 9), Los Angeles on the west side, San Francisco on the east. Horizontal movement to the right along the fault shifts Los Angeles northward and San Francisco southward. Straddle the fault at San Francisco, looking south, to visualize the rock mass carrying Los Angeles on your right moving toward you. You can also look across the fault at San Francisco and imagine that the rock mass that carries the city is moving to your right. Sudden jerks of rock masses along the fault set off frequent earthquakes which testify that the movement is ongoing.

Average movement along the fault is less than an inch per year. At this rate, the two cities could be suburbs of one another in about 25 million years! Not an event to be concerned about for awhile.

Q 10: A geologist friend told me of another idea which also seems wild to me. He says that, in time, Billings, in south-central Montana, will sit among the hot springs and geysers of what was once Yellowstone National Park. A myth, right?

A: Not necessarily. Recall the concept of plate tectonics (Chapter 20). Part of the concept accepts the idea of *hot spots*, isolated places where rising plumes of molten rock burn to the surface or have done so in the recent past. The plumes may rise hundreds of miles (kilometers). Many hot spots have been speculated upon.

116 Geology to Stir the Brain

Among them are those at Iceland, Hawaii, and Yellowstone National Park. Fairly recent volcanic activity, hot springs, and geysers support the concept of a hot spot at Yellowstone.

Lavas in the Snake River Plain of southeastern Idaho generally become younger to the northeast with the youngest in Yellowstone. An explanation for the geologic facts is that the North American plate has been moving to the southwest over a hot spot in the southeastern Idaho-Yellowstone Park region during the past 2 million years or so. The track of lava emplacement in Idaho and Yellowstone can be extended northeast to the vicinity of Billings. So if the movement of the plate remains constant, as well as the position of the supposed hot spot, Billings could, one day, find itself amongst lava flows, hot springs, and geysers. But this might not happen for hundreds of thousands or even millions of years.

Earth History
Q 11: Did cave men ever slay a dinosaur?

A: No. The last dinosaurs died out permanently--became extinct--65 million years ago. The oldest humans that stood upright and walked on two feet can be traced back to only about 4 ½ million years. So dinosaurs and humans did not co-exist, in spite of the cartoon that depicts congenial Alley Oop riding on the backs of dinosaurs.

Q 12: What caused the extinctions of the dinosaurs?

A: If you knew the verifiable answer to that question, you'd be famous. A much-touted theory says a meteorite struck Earth 65 million years ago to cause dinosaurs' extinction. A huge amount of dust and smoke was thrown into the atmosphere, blocking sunlight and lowering global temperature for a year or more. Diminished sunlight retarded plant growth whose shortcoming rippled through the food chain. Reduced temperatures led to what's called an "impact winter" which may have had negative impact on the dinosaurs in its own right. A major objection to this theory for dinosaurs' demise is that these animals had declined well before the time of the supposed meteorite impact.

Another theory involves a large increase in volcanic eruptions over a period of thousands of years. Volcanic ash and dust in the atmosphere would produce impact-winter-like effects along with acid rain and the depletion of ozone.

Dozens of other theories are out there. Some paleontologists look to non-geologic explanations. Among them are widespread disease and the destruction of dinosaur eggs by small, but secretive and brainy mammals. Based on the answer to the previous question, you know humans had nothing to do with dinosaur extinction.

Q 13: Is extinction a fact of life?
A: Yes. Extinctions have happened many times in the geologic past. From fossils, paleontologists know of major extinctions at or near the end of the Ordovician Period (440 million years ago) (Table

Figure 21-1. Trilobite. Age: Cambrian. Jince, Bohemia. The three-lobed fossil, about 4 inches (10 centimeters) long, has its head region to the left.

20-1), Devonian (365 million years ago), Permian (250 million years ago), Triassic (215 million years ago), and Cretaceous (65 million years ago). The times in years are approximate.

Extinctions during the Permian and Cretaceous are most significant. Extinctions in the Permian decimated 85 percent or more of the life then in existence. Marine invertebrates were hardest hit, followed by land animals. The marine trilobites (Figure 21-1), indirectly related to crabs and lobsters, is one notable group that became extinct at the end of the Permian. At or near the end of the Cretaceous, up to 75 percent of the marine creatures died out, followed by land animals, including the dinosaurs.

Extinction continues to claim species today. In the U.S. alone, about 500 species have become extinct since the 1500s, most because of human activity-the destruction of natural habitat, pollution, and the like. Should we be concerned? Yes! Our lives are much enriched by interaction with non-human life. We cherish visits to zoos, arboretums, and state and national parks, marvel at the animals and plants in such places and elsewhere.

Some who don't cherish the diversity and aesthetics of non-human life harbor a what's-in-it-for-me attitude. Are the selfish needs of humans satiated by wild plants and animals? Today, 70 percent of the drugs used in modern medicine derive from plants. Thousands of plant species have not been screened for their medicinal uses. Do we want them extinct before we've had a chance to analyze them for their possible benefits?

But not only plants provide medicinal uses. From the skins of certain frogs, for example, can be extracted drugs for human skin diseases, pain, and other uses. Again, can the skeptics turn their cheeks at the imminent extinction of creatures that might make human life healthier and less demanding medically?

Q 14: I've heard that North America was covered by seas many times. How is this so?
A: You must go to the rocks and the fossils they contain. Many fossils, although extinct, belong to groups that live today only in salt water such as most brachiopods (BRACK-ee-uh-podz, Figure 22-1). Brachiopods are shellfish with two half shells, as for clams, but are unrelated to clams. If we find such fossils in sandstones, shales, and limestones, we can safely assume that the fossils and these rocks which contain them were laid down in seas. If we find marine fossils and rocks stacked up in many layers and can age the rocks with the fossils, we can say with confidence that seas covered North America many times.

Q 15: Where did Earth's ocean water come from?
A: An earlier theory says ocean water has come from Earth's interior by a degassing process brought about by a hot, molten condition at depth. Hot clouds of water vapor eventually condensed to rainwater to form the oceans as Earth cooled. Why is this a plausible theory? Because today steam makes up 70 percent or more of the gases emitted from volcanoes, and degassing continues to the present day each time a volcano erupts.

Another theory is that impacting comets may have provided most of Earth's ocean water billions of years ago. Evidence to support this theory comes from Comet LINEAR, which broke apart in August

2000 as it passed the sun. During its destruction, scientists say that the comet contained water with the same composition as water on Earth.

Q 16: I thought badlands are badlands. But those of Badlands National Park in South Dakota and Theodore Roosevelt National Park in North Dakota look different. Why?

A: Early French explorers and Native Americans considered badlands "bad" because of the difficulty of crossing them. These easily eroded terrains, usually with little plant cover, are cut up easily by streams into a maze of narrow ravines, sharp ridges, and pinnacles. Some badlands display many rounded slopes if the rocks are mostly of finer grained claystone (lithified clay) and siltstone (lithified silt).

The badlands of the two parks look different because they differ in the types and ages of the rocks. In South Dakota, pastel grayish–pink and tan shades of the rock layers result from stream and lake sediments, volcanic ash, and old soils laid down mostly about 20 to 35 million years ago.

Generally brighter shades of yellow, light gray, and blue in the North Dakota badlands also result largely from older stream and lake sediments and volcanic ash, deposited 55 to 60 million years ago. But what sets these badlands apart are the layers of black lignite coal and the pink to red and orange "scoria." This scoria differs from true scoria, a volcanic rock, in being formed by burning lignite baking overlying layers of sediment. Small amounts of iron stain the baked sediments reddish.

If possible, try to enjoy the rock colors of either of the badlands after a rain. At this time, the hues saturate.

Q 17: Will Niagara Falls last forever?

A: I'm afraid nothing on Earth is forever, especially waterfalls. Niagara Falls occur on the northwest edge of New York State. They came into being about 8,000 years ago when the continental ice sheet in the Great Lakes area melted back far enough so the north–flowing Niagara River could drain from Lake Erie to Lake Ontario. Niagara River water originally dropped over a line of Niagara cliffs that face northward. Resistant dolostone that holds up the cliffs is underlain by mostly weak shales.

Streams tend to erode toward their upstream or headward ends (Chapter 2). In like manner, waterfalls and rapids migrate upstream. In the case of Niagara Falls, turbulent falling water easily erodes the weak shales and undercuts the overlying dolostone which topples into the river. This allows the falls to steadily migrate upstream at a rate of 4.3 feet (1.3 meters) per year. The falls have retreated 7 miles (11.2 kilometers) since their origin. They will become obliterated as they wind their way to Lake Erie.

One day honeymooners will no longer be able to revel in Niagara Falls' thunderous beauty. The falls will be relegated to the dustbin of Earth's memorable features.

Q 18: How can geologists work out climates of the past?
A: Mostly from certain rocks and fossils which give direct clues to climate. Thick and widespread limestones with fossil corals and other reef-forming animals are good indicators of tropical to subtropical climates. Today, limy sea bottoms are found in such places as the tropical Bahamas and Persian Gulf. Limy sea bottoms and reefs (along with salty rocks, see below) existed, for example, over much of North America about 400 million years ago, during the Silurian Period (Table 20-1). We can, therefore, infer a tropical to subtropical climate on the continent during that time.

Rocks like rock gypsum and rock salt form by evaporation from brines. An arid climate is necessary for evaporation to occur on a large scale and create great thicknesses of these rocks.

Thick layers of lithified sand dunes also point to an arid climate. And internal layering of the dunes parallels the downwind slopes (Chapter 4). Such layering recognized in lithified dunes documents wind directions in the past.

Widespread and thick deposits of coal indicate a humid climate. But past temperatures may have varied from cool to tropical.

Petrified wood with well-defined growth rings verify a temperate climate within which the trees grew. Such rings tend not to develop in tree trunks grown within a tropical climate.

A mixed rock of clay to boulders can be an indicator of having been deposited by glaciers in a moraine (Chapter 7) within a cool, moist climate. This mixed rock is best confirmed if it overlies a grooved and scratched rock surface. Some such mixed rocks can be deposited by mudflows and other means, so past glacier sediment must be inferred with care.

Q 19: How do paleontologists use fossils to date rocks?
A: Paleontologists, those who study fossils, deal with vertebrates, invertebrates, plants, and tiny microfossils, which may be animals, plants, or other forms of life. Life has changed in a natural progression through time. The theory of evolution is significant in understanding this progression. Once a certain life form changes, it doesn't revert back to its original form. Life has not merely changed

through time, but has increased its complexity, and much has become extinct. Knowing the history of life through time-including the extinctions--and being able to identify fossils in detail, paleontologists can assign relative ages to fossils and rocks:

Let's see how dating of rocks by fossils might work. Assume you examine a road-cut exposure of sandstone that breaks down into loose sand in places. You spot a tooth-like fossil (Figure 21-2), pick it up, and warm with the thrill of discovery. It's beautiful! Might be a

Figure 21-2. Shark tooth. Age: Late Paleocene. Southwestern North Dakota. The tooth is 0.45 inch (11.4 millimeters) high.

shark tooth. You show it to a paleontologist friend who shares your excitement.

"You're right on. It's one of the side teeth of a shark."

Your grin widens.

He checks the fossil collection where he works, consults a few publications."I'm able to identify it to species: *Paraorthacodus eocaenus* (pear-uh-ore-thuh-COE-duhs ee-uh-SEE-nuhs)."

The technical name flows with such ease from his mouth, and his smile seems to indicate his pleasure in casting such names about.

"The two pairs of tiny teeth on either side of the main tooth give it away. This species has been reported from England, Belgium, France, western Greenland, and Russia, nearly all finds from the late Paleocene (Table 20-1). It's a good age indicator. Congratulations."

Q 20: Why do paleontologists give fossils such long, hard-to-pronounce names?

A: I assume you refer to so-called scientific names set off in italics, like that of the fossil shark above, or like *Tyrannosaurus* (tuh-ran-uh-SOAR-uhs), the name of a Cretaceous dinosaur. In actuality, paleontologists share the coining of these names with biologists.

Before I answer your question, let me digress a bit. Scientific names are a kind of international language. They follow a set of rules that are accepted world-wide. A paleontologist from India can communicate with one from Canada, at least where fossil names are concerned.

Scientific names also eliminate confusion. A fossil may have several common or vernacular names in various countries, but rules permit only one scientific name. To strengthen the point, in English-speaking countries, not enough common English names are available to accommodate both living and fossil species.

You may have noticed that each complete name is made up of two names, a generic name, with the initial letter capitalized, and a specific name, with the initial letter not capitalized. So the complete name of the Cretaceous dinosaur is *Tyrannosaurus rex.*

Names created well say something about the species. *Tyrannosaurus* comes from the Greek *tyrannos*, tyrant and *saurus*, lizard. The specific name is from the Latin *rex* meaning king. So the full name can be translated as "King of the Tyrant Lizards," which has been considered a ferocious carnivorous dinosaur. Some now believe the beast may have been a scavenger.

Back to your question. Not all names of fossils, and those of living organisms as well, are long. Consider the snail *Turbo* and the clams *Venus* and *Mya*..

Chapter 22
Geological Puzzles: A Selection

Many people like to work out crossword puzzles. With your geological background from this book, you might want to try your hand at several geological puzzles, worked out from photographs. Try to reason out the puzzle before you check my explanation. After all, if you're wrong, no one should slap your hand!

We'll begin with relatively simple challenges and work toward more involved ones. Puzzles are grouped according to those visible on the ground and those seen from the air.

From the Ground
Puzzle 1 (Figure 9-1). **Rock layers and San Juan River. Southeastern Utah. What happened here?**

Notice that the rock layers at the upper left of the photograph are nearly flat-lying. As you trace the distant layers to the right, they bend downward at a sharp angle. The layers in the mid-distance, however, are horizontal as are those in the immediate foreground.

All these rock layers were buried by other rock layers at one time. From what we can see here, Earth forces squeezed the layers into a single massive bend or fold that was possible when the rock layers were at some depth and under considerable pressure. Erosion wore away the overlying rock layers to expose the fold. The San Juan River cuts through all the rock layers today.

Puzzle 2 (Figure 9-2). **Tilted rock layers overlain by flat-lying layers. Grand Canyon National Park, Arizona. What happened here?**

The tilted rock layers, extending from the river to near the top of the photograph, were once flat-lying or nearly so as are most sediments deposited today. Earth forces squeezed the rocks and tilted them to the right. Erosion planed off the edges of the tilted strata for a time before near-horizontal rock layers buried them. Geologists call the buried surface of erosion that separates the tilted and flat-lying strata an *unconformity*. This surface relates to time not represented by a record in the rocks. Other such surfaces are present within the sequence of rock layers in the Grand Canyon. After all the rocks seen here and others that overlie them were laid down, the Colorado River cut the Grand Canyon (Chapter 21, Question 2).

Figure 22-1. Fossil shells in rock slab. Age: Mississippian. Burbank, northeastern Ohio. The largest shell in the upper right is 0.87 inch (22 millimeters long). **Puzzle 3.**

Puzzle 3 (Figure 22-1). Fossils in rock slab. Northeastern Ohio. What happened here?

This slab of Mississippian-age siltstone contains numerous, elongate clams and nearly circular brachiopods. Most of the fossils are only impressions but many of the clams contain remnants of shells that haven't yet been dissolved away. Comparisons with living relatives confirms that the clams and brachiopods once lived in a sea.

Nearly all the shells lie with their convex surfaces uppermost, a stable position. Many of the clams cluster or generally align toward the top of the photograph. These positions suggest that waves or currents oriented the shells somewhat before they were covered by sediments and preserved as fossils. Associations of fossils like these give notable insight into past communities and environments

Figure 22-2. Rod-like fossils in whitish sandstone. Age: Cretaceous. Southwest of Rhame, southwestern North Dakota. **Puzzle 4.**

Puzzle 4 (Figure 22-2). Rod-like fossils in whitish sandstone. Southwestern North Dakota. What happened here?

This may be a tough one. You are looking at rusty-weathering (oxidized) sandstone rods that lie parallel to the layering of the whitish sandstone. The rods are about 1 to 1 ½ inches (2.5 to 3.8 centimeters) across and covered with tiny bumps or knobs. You can see that the rods branch. Odd?

Paleontologists refer to indirect evidences of life as trace fossils, such things as tracks, trails, borings, and burrows. In other words, no direct evidence of the body of an organism exists, only an indirect trace.

At question here, are the filled burrows of a marine organism. Today, in shallow seawater, certain ghost shrimp and similar creatures construct burrows in sand with knobby outer surfaces. With their claws they fashion pellets of sand mixed with mucous and press them against the burrow wall in brick-layer fashion. Such work can create a burrow with a knobby exterior. A similar creature back in the late Cretaceous (Table 20-1) probably produced the burrows seen here. The original burrows were later filled with sand and cemented with mineral matter, making the burrows more resistant to weathering and erosion than the enclosing whitish sandstone. The technical name given to these burrows is *Ophiomorpha* (oh-fee-uh-MORE-fuh).

Years ago, I spotted fossil burrows much like these within a glass display case in a small-town museum. A scrawled label next to them read, "Fossil Corncobs." I considered correcting this error with the museum's caretaker, but decided against it. The fossils were already identified. Why listen to a smart-aleck stranger presenting himself as an expert?

From the Air

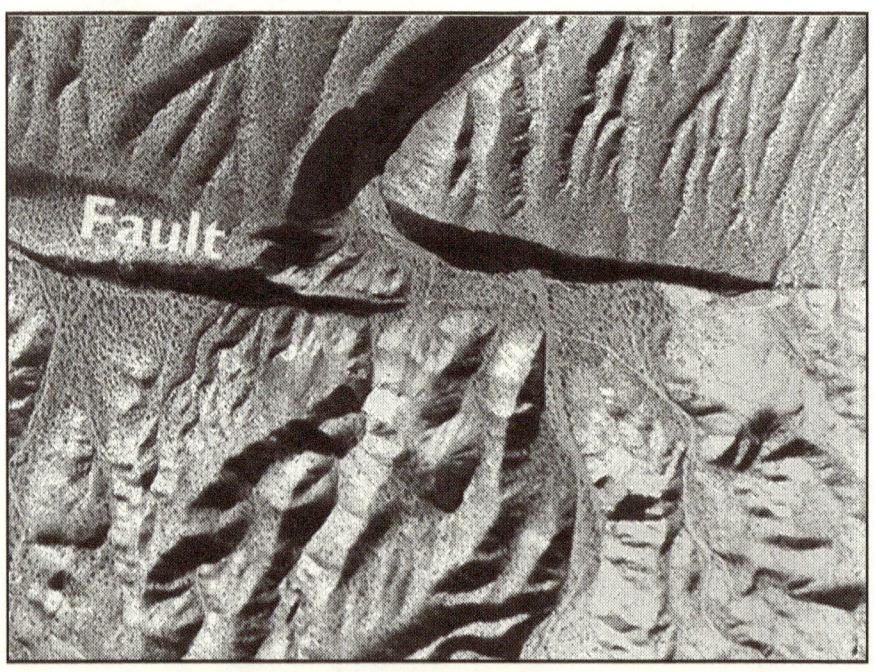

Figure 22-3. Offset of stream drainages. El Paso Mountains, west of Searles Valley, San Bernardino County, southern California. Photograph by U.S. Geological Survey. **Puzzle 5.**

Puzzle 5 (Figure 22-3). Offset of stream drainages. Southern

California. What happened here?

In the upper part of the photograph, a broken line, roughly from left to right, divides the landscape sculpted by intermittent streams. The broken line is the trace of the Garlock Fault, a northeast-trending branch off of the San Andreas Fault, along which side-slip movement of rock masses has taken place. This side-slip movement, very recent, has offset the stream valleys. Notice especially that the major valley in the upper central part doesn't have a true counterpart below the line of division.

Figure 22-4. Linear features. Confluence of Colorado and Green Rivers, Canyonlands National Park, southwest of Moab, southeastern Utah. Photograph by S.W. Lohman, U.S. Geological Survey. **Puzzle 6.**

Puzzle 6 (Figure 22-4).   Linear features and river confluence. Canyonlands National Park, southeastern Utah.   What happened here?

North is toward the top of the photograph. At the upper left is the confluence of the Colorado and Green Rivers–where they join. The Colorado enters the photograph at the top, and flows southwest. The Green enters at the upper left.

The four major north–northeast–trending linear features are grabens (Chapter 9), down–dropped rock blocks that now form vertically–walled rift valleys.   They occur within an appropriately named southern region of the park called The Grabens, southeast of the Colorado River. The grabens are the result of pulling–apart Earth movements at right angles to the grabens. Look closely and you will see numerous rock fractures that parallel the grabens. Still other cross–fractures trend to the northwest.

Puzzle 7 (Figure 22-5). Sheep Mountain. North-central Wyoming. What happened here?

Sheep Mountain, just northwest of Greybull, Wyoming, trends northwest.   Concentric bands of eroded rock layers tell us the mountain is a massive fold. Rock layers that tilt away from the crest of the mountain on both sides further tell us that we are dealing with an up–buckled, deeply eroded anticline.   Layers to the northwest outline a spearhead–like pattern to reveal that the whole anticline is tilted in that direction. Although not visible, the fold tilts as well to the southeast. This is a classic example of such a rock fold, best appreciated from the air.

Eroded anticlines display older rocks near their centers and younger rocks away from them. For Sheep Mountain, the oldest rocks exposed where the Bighorn River cuts through it are Mississippian in age (Table 20-1).   Rock layers are Cretaceous in age on the outer flanks of the mountain.

This anticline had to have formed at considerable depth where rocks bend rather than break.   Much erosion has brought it to the surface.

The Bighorn River strangely, so it seems at first, cuts right through the resistant rocks of the anticline. But Question 3, Chapter 21, covers why some streams cut through mountains rather than go around them.

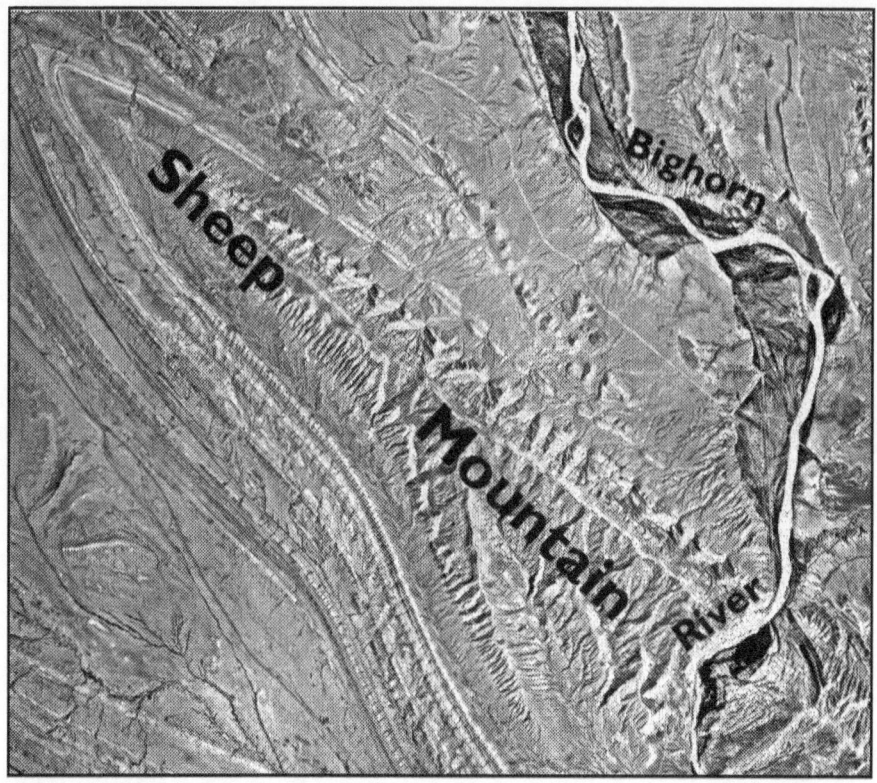

Figure 22-5. Sheep Mountain. Northwest of Greybull, north-central Wyoming. The Bighorn River cuts through the mountain. North is toward the top of the photograph. Aerial photograph 7801-139 by the U.S. Geological Survey. **Puzzle 7.**

Puzzle 8 (Figures 22-6 and 22-7). Ship Rock, Northwestern New Mexico. What happened here?

Ship Rock is a mountain peak in northwestern New Mexico, southwest of the town of Shiprock. It projects 1,700 feet (518 meters) above the New Mexico plain, like a majestic volcanic ship in a sea of shale. Three main, sharp-crested, wall-like ridges extend out from it.

Thirty to 19 million years ago an explosive volcano erupted over what is now Ship Rock. After eruption ceased, molten rock cooled and solidified in the throat of the volcano, together with lithified volcanic rock fragments. Over time, weathering and erosion stripped the cover of less resistant rock away, leaving the more resistant rock of the volcano's throat. Ship Rock, today, represents

Figure 22-6. Ship Rock. Southwest of Shiprock, northwestern New Mexico. A wall-like dike leads to Ship Rock. Photograph 3063 by W.T. Lee, U.S. Geological Survey. **Puzzle 8.**

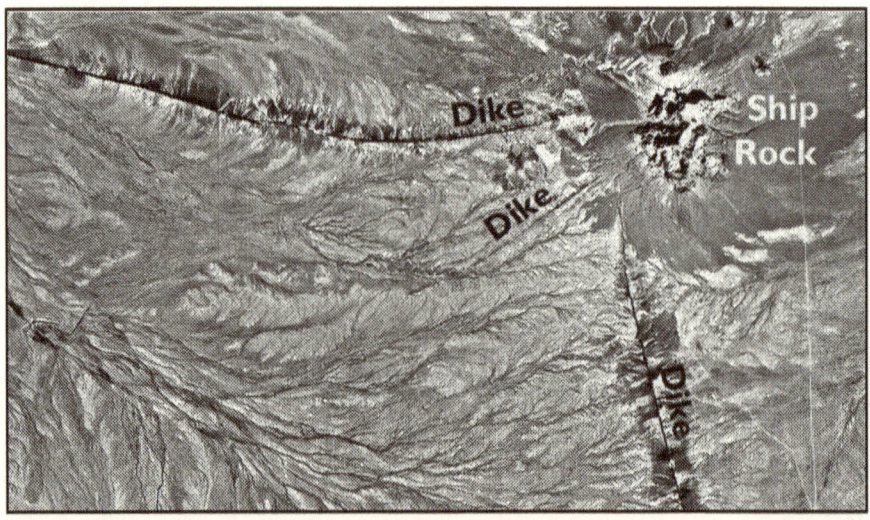

Figure 22-7. Ship Rock, aerial view. From the air, three main dikes are visible. North is toward the top of the photograph. Aerial photograph 9959-35 by U.S. Geological Survey. **Puzzle 8.**

a volcanic neck (Chapter 5), only a remnant of a former volcano.

The wall-like ridges joined to Ship Rock? These are eroded dikes (Chapter 5), emplaced into huge vertical cracks at the time that the Ship Rock volcano came into being. In the aerial view (Figure 22–6) three main dikes, of dark volcanic rock, are visible: to the south, west, and southwest. The dike to the southwest, which extends partly to the northeast, appears as a fine line, only 7.5 feet (2.2 meters) wide.

Chapter 23
How Roads Reflect Geology

We can read some geology just from a casual glance at a highway map. The connection has to do largely with mountains or the lack of them. Roads, of course, cut across mountains, but contractors build the vast majority of roads around them or within their valleys.

Look at the patterns of roads for your road-geology read. The patterns reflect the underlying geology. I've assembled examples from five places in the United States to show you how the process can work.

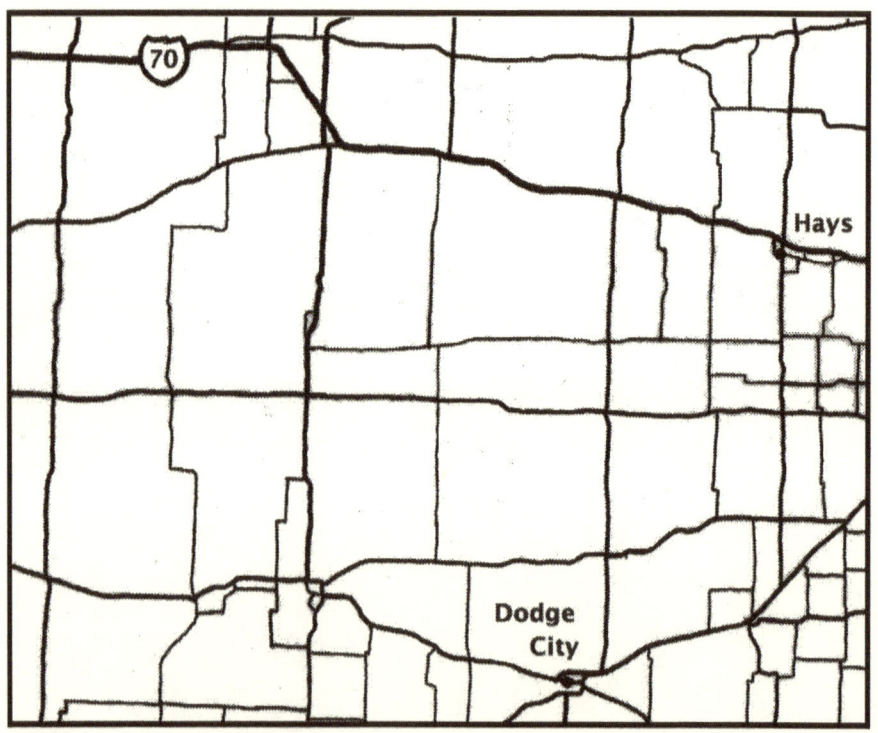

Figure 23-1. Major roads in west-central Kansas. The air distance between Dodge City and Hays is 87 miles (140 kilometers). North is toward the top of the figure.

West-central Kansas (Figure 23-1)

Most of the roads create a rectangular pattern. They tend to run north–south or east–west, and tend to parallel the boundaries of legal land divisions (township, range, section). Some roads are slightly crooked where they follow stream valleys. Most of the road west-northwest of Dodge City follows the Arkansas River Valley as does part of the road northeast of that city.

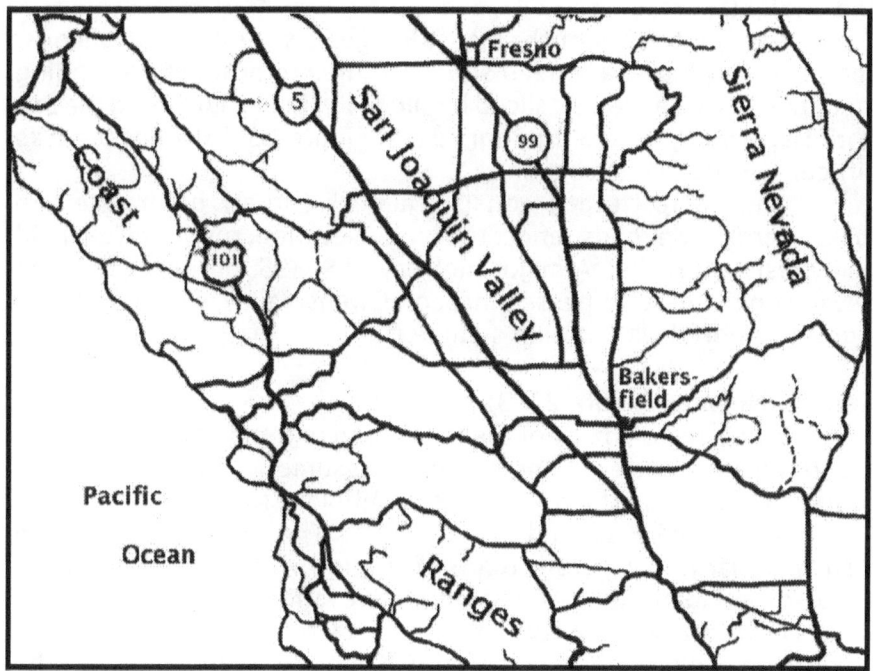

Figure 23-2. Major roads in south-central California. Only a selection of minor roads is shown in the Coast Ranges and Sierra Nevada. The air distance between Fresno and Bakersfield is 103 miles (166 kilometers). North is toward the top of the figure.

Streams in this part of Kansas course eastward over shale, sandstone, and limestone which are not particularly resistant to erosion. The strata are essentially flat-lying or bent downward only slightly into a slight basin. In essence, the roads are little affected by the geology or not at all--the simplest relationship.

South-central California (Figure 23-2)

This region shows three kinds of road patterns. In the San Joaquin Valley, the major highways California 99 and Interstate 5 are relatively straight and trend northwest. In the Coast Ranges, U.S. 101

also trends northwest but is twisted. The roads in the Sierra Nevada are nearly all minor and show an irregular pattern at best.

The San Joaquin Valley, a southern extension of the Great Valley, is both a surface basin and one at depth with the rock layers bent downward. A flat valley floor surfaced with stream sediment offers ease of road construction, so the majority of the roads are long and straight.

Resistant rocks of the Coast Ranges--sedimentary, igneous, and metamorphic--have been squeezed up into northwest-trending mountains. These are also sliced by slide-slip faults, including the San Andreas. Highly crooked minor roads (Figure 23-2) suggest rocks difficult for road construction.

The Sierra Nevada consists almost entirely of granite-like rocks. Serpentine roads reflect the high-elevation, rough terrain held up by resistant rocks. A major highway, U.S. 395, borders the Sierra Nevada on the east. It passes through Owens Valley, which occupies a graben floored with stream sediment.

Central Nevada (Figure 23-3)

In this region many of the roads tend to trend north-south and are irregular in pattern. Interstate 80 assumes a somewhat snaky course because it partly follows the Humboldt River Valley.

This is the Basin and Range region. The strange road pattern results, in large part, from roads that align with valleys along the edges of elongate, intervening mountain ranges. The flat-floored valleys are actually rift valleys covered with stream sediment. And the ranges are fault-block mountains (Chapter 9), held up mostly by resistant sandstones, limestones, and volcanic rocks. On a drive west from Ely on U.S. 50, "The Loneliest Road in America," you ride up and down over numerous ranges and basins, all the result of faulting.

Southwest-central Pennsylvania (Figure 23-4)

Two road patterns are evident here. One, within the Appalachian Mountains, has a definite "grain" and trends northeast as it bends easterly. State College is within this pattern. The other, within the Allegheny Plateau, contrasts in its lack of any definite "grain." Johnstown lies within this pattern, whereas Altoona rests at the boundary between the two regions.

The road pattern in the southeast reflects the Valley and Ridge region, the western part of the Appalachian Mountains, an area of tightly folded anticlines and synclines (Chapter 9). Squeezing for their

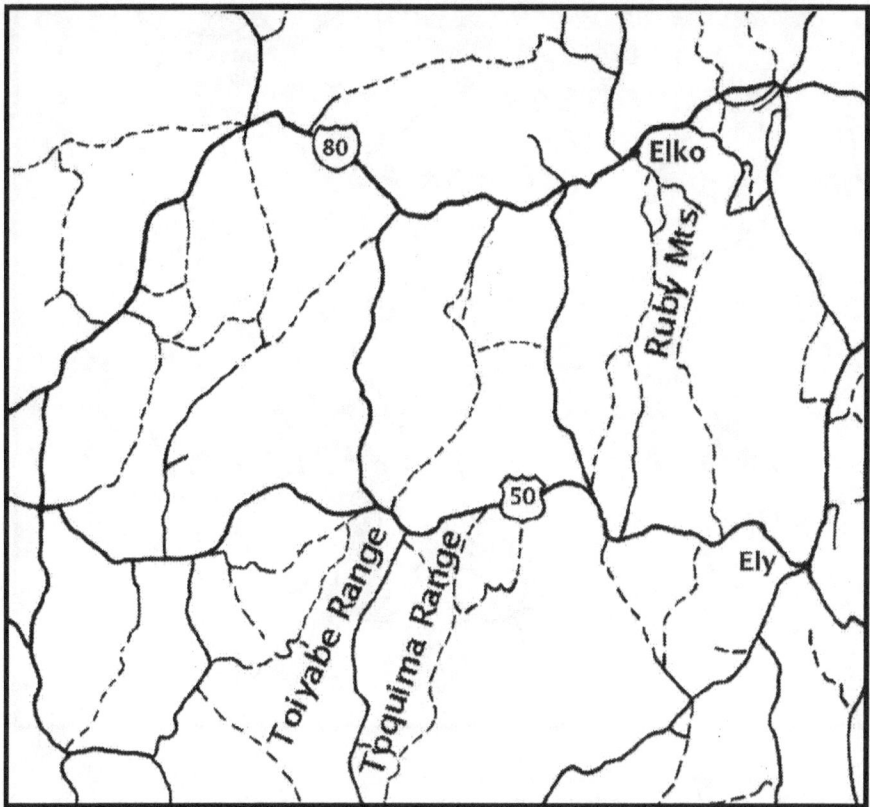

Figure 23-3. Major roads in central Nevada. Roads shown as dashed lines are unpaved. The air distance between Elko and Ely is 111 miles (179 kilometers). Only three of the mountain ranges are labeled. North is toward the top of the figure.

formation came from the southeast. As the folds were worn through, resistant sandstones and some limestones formed the ridges-actually low mountains-and easily eroded shales formed the valleys through which many of the roads pass.

Rock layers in the Allegheny Plateau in comparison are only gently folded, and made up mostly of sandstone and shale. This combination gives no obvious pattern to the roads.

Black Hills, South Dakota and Wyoming (Figure 23-5)

Let's step back a bit for this one. The Black Hills, although most often allied with South Dakota, occur also partly in Wyoming. These so-called hills are an uplift of moderate mountainous terrain,

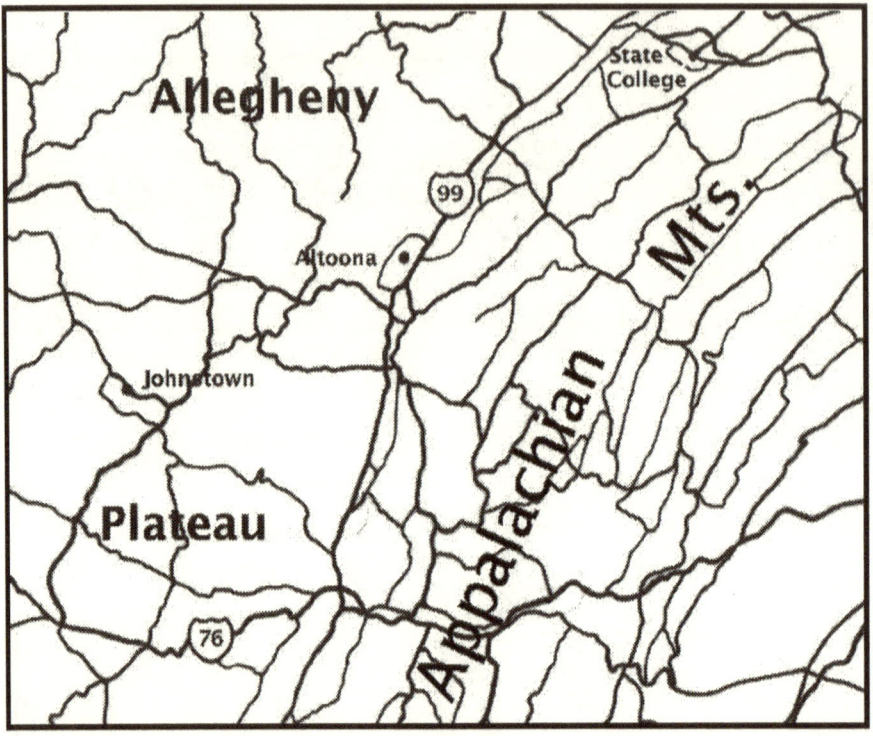

Figure 23-4. Major roads in southwest-central Pennsylvania. Altoona is on the boundary between the Allegheny Plateau and Appalachian Mountains. The air distance between Johnstown and State College is 64 miles (103 kilometers). North is toward the top of the figure.

about 170 by 70 miles (274 by 113 kilometers), that rises above the surrounding Great Plains. The uplift began as a gigantic Earth blister or dome, formed by squeezing forces some 60 to 65 million years ago. As erosion took its toll, oldest rocks were exposed near the center--as old as 2.5 billion years--with progressively younger rocks in concentric bands toward the periphery. Some of the upturned edges of Cretaceous strata form conspicuous hogbacks.

Interstate 90 and South Dakota 79 conform closely with the eastern configuration of the Black Hills dome. Between Rapid City and Spearfish, I 90 follows the Red Valley made up of red and green shales. Sandstone hogbacks (Figure 9-3) closely margin the Red Valley on the northeast. South Dakota 79 mostly courses on gray shales between Rapid City and Hot Springs, with sandstone hogbacks just to the west. Similar hogbacks arise also in the southwestern part

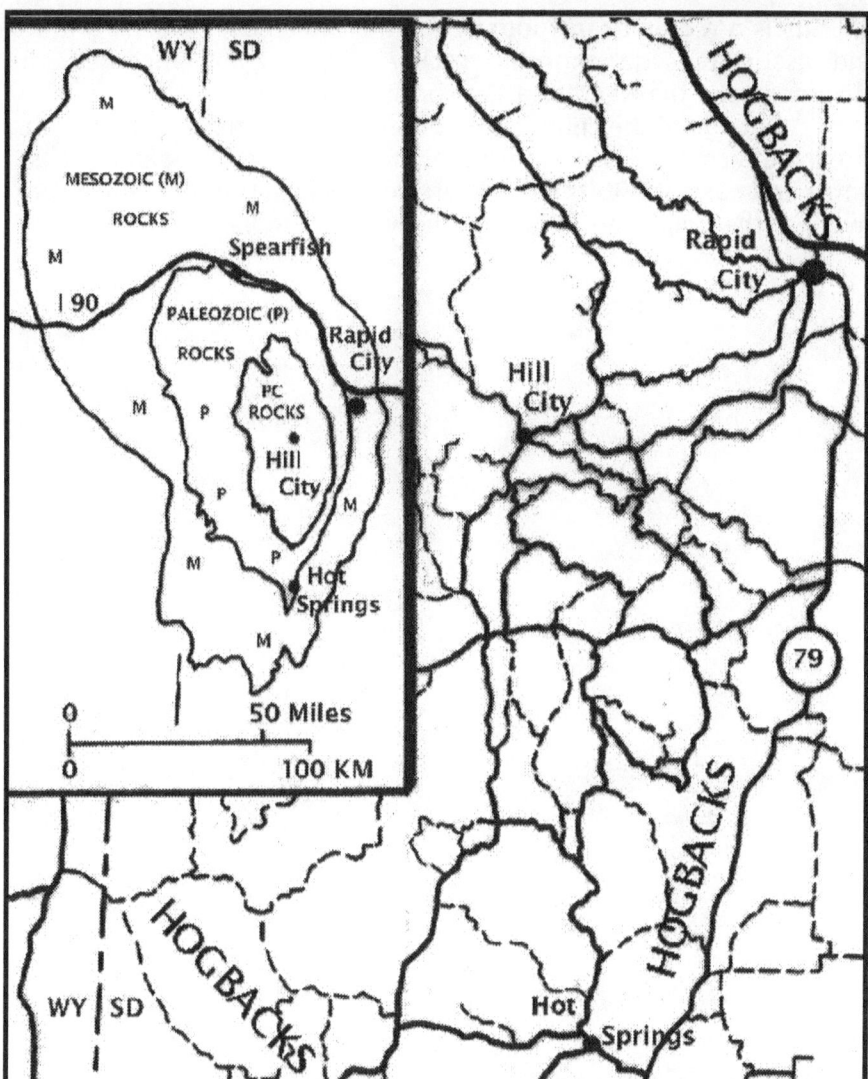

Figure 23-5. Major roads in central part of Black Hills, South Dakota and Wyoming. The inset shows a simplified geologic map of the Black Hills uplift; PC=Precambrian. The air distance between Spearfish (inset) and Hot Springs is 83 miles (134 kilometers). North is toward the top of the figure. Interstate 90 runs northwest from Rapid City.

of the South Dakota part of the Hills.

Hill City centers the exposed core of oldest rocks--Precambrian--in the Hills, made up of granite and various metamorphic rocks: schist, gneiss, marble, and the like. Southeast of

Hill City is a region of tortuous roads that travel over mostly granite and associated metamorphic rocks. This region includes Mt. Rushmore and the Needles areas, both in granite.

Outside of the central core of oldest rocks, the irregular roads mostly reflect travel over limestone and dolostone (Paleozoic rocks). North and east of I 90 the roads are relatively straight because rock layers of the uplift here tilt at relatively slight angles.

Glossary of Terms in Geology
Used in This Book

aa A Hawaiian term for a kind of lava flow with a rough, jagged surface.

aftershock A smaller vibration that follows a main earthquake.

alkali Any of several white, bitter-tasting salts found at the surface in dry regions.

alluvial fan A fan-shaped pile of sediment laid down by a stream, usually where it leaves a mountain range and enters a flat plain.

anticline An up-buckled or up-arched fold of rock layers.

aquifer An underground rock layer, such as sandstone, that bears water.

asteroid A rocky or metallic body smaller than a planet. Most orbit about the sun in a belt between Mars and Jupiter.

bar A ridge of sand or gravel at the bottom of a stream, lake, or shallow sea.

barchan A crescent-shaped sand dune with its pointed ends directed downwind.

barrier island An elongate island of sand or gravel that parallels a seacoast or lake coast, and is separated from a coast by a lagoon.

basalt A dark, fine-grained igneous rock, which makes up most of the oceanic crust.

basin 1. A depressed area into which streams drain. 2. A circular or elliptical syncline or down-fold of rock layers.

baymouth bar A ridge of sand or gravel that blocks the mouth of a bay.

beach drifting The movement of sediment parallel to a beach by waves striking the beach at an angle.

blowout A depression scooped out in sand by wind.

brachiopod A mostly marine animal with two limy half shells of different size and shape; unrelated to clams.

butte An isolated hill, often with steep sides, in arid regions; smaller than a mesa.

calcite A common calcium carbonate mineral, the main constituent of limestone.

caldera A circular or ellipse-shaped depression at the top of a volcano, larger than a crater and more than about a mile (1.6 kilometers) across.

cinder cone A small volcano composed of loose volcanic rock fragments such as cinders.

cirque A half-bowl-like depression at the upper end of a valley where a valley glacier originates.

columnar structure A structure in volcanic rock that resembles closely packed columns or posts, mostly six-sided; also called columnar jointing, with emphasis on the fractures that outline the columns.

comet An object in space made up of rock and ice.

composite volcano A usually large volcano composed of alternating layers of lava and volcanic rock fragments.

conglomerate The lithified equivalent of gravel, of rounded pebbles up to boulders.

continental drift An early theory that continents have drifted through ocean basins; replaced by the plate tectonics concept.

continental glacier A glacier ice sheet that covers a large part of a continent.

core The central part of Earth's interior.

crater A circular or ellipse-shaped depression at the top of a volcano, smaller than a caldera and less than about a mile (1.6 kilometers) across.

creep The imperceptible slow movement of soil and rock down a slope.

crust The outermost zone or layer of Earth.

cuesta A ridge held up by a resistant, tilted rock layer. One slope is steep, the other gradual.

debris slide A kind of landslide in which rock debris slides rapidly downslope.

debris fall A kind of landslide in which rock debris falls from a cliff.

debris flow A kind of landslide in which rock debris, mud, and water flows rapidly downslope.

delta A body of sediment laid down at the mouth of a stream where its speed is checked.

desert pavement A layer of packed pebbles left behind in a wind blowout in arid regions.

dike A flat-sided body of rock that cuts across other rocks.

distributary A channel split off from a main stream channel at a delta; the opposite of a tributary.

dolomite A calcium magnesium carbonate mineral that makes up dolostone.

dolostone A limy rock like limestone but made up of the mineral dolomite.

dome 1. A raised area from which streams drain. 2. A circular or ellipse-shaped anticline or up-fold of rock layers.

downcutting Erosion of a stream bed.

dripstone A deposit, like a stalactite or stalagmite, formed by dripping water in caves.

drumlin A streamlined, elongate hill parallel to the flow of a former glacier.

earthquake A trembling of the ground caused by sudden release of energy caused by faulting or volcanic activity.

epicenter A point on the surface above the source of an earthquake.

erosion The loosening of any Earth material, and its movement by water, wind, ice, and gravity; includes weathering.

erosional mountain High-elevation mesa or plateau held up by rocks resistant to erosion.

esker A snaky ridge of sand and gravel which represents a stream that flowed on, within, or beneath a glacier, and was left behind after the glacier melted.

estuary A drowned or partly submerged stream mouth along a seacoast.

fault A break in rocks along which rock masses have shifted.

fault-block mountain A mountain formed by pulling-apart forces and faulting on both sides of a mountain block. Blocks on either side of the mountain drop down to form elongate basins, also called rift valleys.

feldspar A white, gray, or pinkish aluminum mineral rich in potassium, calcium, or sodium. It breaks into tiny blocks, and occurs in many kinds of rocks.

flash flood A sudden, but short-lived, flood triggered by intense thunderstorms.

floodplain The flat area adjacent to a stream covered by water during a flood.

flowstone A deposit formed by flowing water in a cave.

folded mountain A mountain made up of folded, and also usually faulted, rock layers.

foreshock A smaller vibration that precedes a main earthquake.

fossil Any evidence of past life.

geologic time Prehistoric time since Earth's origin.

geyser A hot spring that erupts on occasion.

glacier A mass of flowing ice, confined to a mountain valley or not.

graben A down-dropped block of rock, bounded by faults on both sides; revealed as an elongate rift valley.

granite A coarse-grained igneous rock made up of feldspar, quartz, and minor dark minerals such as black mica.

gypsum A soft, clear, white, or gray mineral that can be scratched with a fingernail.

hanging valley A tributary valley whose floor lies or "hangs" above the main valley. It often forms where a main valley is deepened by a valley glacier.

hogback A ridge held up by a resistant, tilted rock layer. Both slopes are relatively steep.

hoodoo A rock pillar eroded into an unusual shape, sometimes resembling a human form.

hot spot An isolated place where a rising plume of molten rock from depth burns through to the surface.

igneous rock A rock formed by cooling and solidifying of molten rock material.

incised meander A loop of a stream cut into underlying rock, most often because of uplift of a region.

joint A crack in a rock along which no obvious movement has taken place.

karst topography A landscape with sinkholes, solution valleys, and no surface streams; results from groundwater dissolving rocks.

kettle A depression formed by the melting of an ice block that was covered by sediment.

landform A feature of Earth's surface with a distinct shape and origin, such as a hill or valley.

lava Molten rock material at the surface; also, the lithified rock material.

lava flow An outpouring of molten lava from a volcanic vent or crack in a rock; also, the lithified outpouring.

lava plain A low-lying tract of land underlain by lava flows.

limestone A rock made up mainly of the mineral calcite.

liquefaction A process whereby water-saturated soil and sediment liquefy during an earthquake.

longshore current A current of water that moves parallel to the shore.

longshore drifting The movement of sediment parallel to the shore by longshore currents.

magnetite A heavy, black mineral attracted to a magnet, and often found in sediment moved by water or wind.

magnitude A measure of an earthquake's strength or the amount of energy released.

mantle The middle zone or layer of Earth's interior, between the crust and the core.

marble Metamorphosed limestone or dolostone.

marine terrace A wave-cut bench along a coast that has been uplifted above sea level.

mesa A high, flat-topped area capped by a resistant rock layer; smaller than a plateau and larger than a butte.

metamorphic rock A rock formed from other kinds of rocks by heat, pressure, and chemical action.

meteor A rocky or metallic body in space that has not struck Earth.

meteorite A rocky or metallic body from space that has struck Earth.

mineral An inorganic naturally formed substance with characteristic physical and chemical properties; such as quartz.

moraine A ridge or uneven landform formed by the laying down of sediment by a glacier.

mudflow A kind of landslide made up of much mud and water.

natural levee A ridge of sand and silt built up on either side of a stream channel after a flood.

outwash plain A nearly flat area of sand and gravel "washed out" beyond a glacier's margin by meltwater and rainwater.

oxbow lake A lake occupying the abandoned loop of a stream.

pahoehoe A Hawaiian term for a lava flow with a smooth, ropy surface.

plateau A high, flat-topped area capped by a resistant rock layer; larger than a mesa.

plate tectonics The concept that Earth's outer shell consists of jigsaw puzzle-like, rigid plates that shift and interact with one another.

playa lake A temporary lake that occupies a shallow desert basin.

pothole A rounded hollow in a rocky stream bed at waterfalls and rapids; formed by the grinding action of sand and gravel.

pumice A rock of frothy, volcanic glass; floats on water.

pyroclastic flow A flow of hot gases, volcanic ash, and rock fragments down the flanks of a volcano.

quartz A glassy mineral that breaks unevenly, and is made up of silica.

quartzite Metamorphosed sandstone.

radioactive dating The dating of rocks based on the decay of radioactive elements.

Richter scale A numerical scale for ranking an earthquake's magnitude.

ripple A small ridge of sand formed by the movement of wind or water; forms at right angles to wind or water movement.

rock A naturally formed consolidated substance made up of one or more minerals; such as granite.

rockfall A kind of landslide whereby rock falls freely.

rockslide A kind of landslide whereby rock slides down a slope.

sea arch An arch in rock formed on a rocky coast by bashing waves.

sea cave A cave in rock formed on a rocky coast by bashing waves.

sediment Loose Earth material--mostly gravel, sand, and mud (silt and clay)--that is moved and laid down by water, wind, ice, or gravity.

sedimentary rock A rock formed from sediment by compaction or by a mineral cement.

seismic wave An energy wave produced by an earthquake or a human-caused explosion.

seismograph An instrument that picks up and records seismic waves.

shale A rock formed by the lithification of mud (silt and clay); breaks up into small chips or plates.

shatter cone A cracked, nested cone in rock; strong evidence for impact by a meteorite.

shield volcano A volcano with gentle slopes and made up mostly of basalt.

silica A natural substance of silicon and oxygen; makes up quartz.

sinkhole A depression formed by collapse of the roof of a cave or cavern or by the dissolving of rocks at the surface; one of the landforms of karst topography.

slot canyon A narrow, straight-sided canyon; dangerous during a flash flood.

soil Loose Earth material formed in-place as rocks break up and decay; a product of weathering.

solution valley A valley formed by the dissolving of rocks; one of the landforms of karst topography.

spatter cone A small, steep-sided volcano of pasty lava.

spit A ridge of sand and gravel along a coast that extends from land into open water.

stack A rocky island along a coast shaped by waves.

stalactite Dripstone that hangs from the ceiling of a cave or cavern.

stalagmite Dripstone that projects upward from the floor of a cave or cavern.

storm surge Short-term rise in sea level brought about by a hurricane.

stream terrace A bench-like remnant of a former floodplain, and higher than the present floodplain.

subsidence The mostly vertical down-sinking of rock and rock debris.

superposition The principle that in an undisturbed sequence of rock layers the oldest lie at the bottom, the youngest are at the top.

surf The zone of breaking waves along a shore.

syncline A down-buckled or down-bent fold of rock layers.

talus A half-cone-like accumulation of rock and rock debris at the base of a cliff.

trilobite An extinct, segmented, marine animal divided into three lobes lengthwise; indirectly related to crabs and lobsters.

tsunami A water wave generated by an earthquake, volcanic eruption, underwater landslide, or meteorite impact.

unconformity A buried surface of erosion; a break in the continuity of a rock record of geologic time.

uniformitarianism The principle that Earth processes operating today have also operated in the past.

valley glacier A glacier confined to a mountain valley.

volcanic mountain A mountain formed by a volcanic eruption.

volcanic neck A butte or mountain that represents the filled throat of a volcano after most of the volcano has been eroded away.

volcano A butte or mountain of lava or volcanic ash and rock fragments.

water table The top of the zone beneath the surface that is saturated with groundwater.

wave-cut bench A bench of rock cut by wave erosion; extends seaward from the base of a sea cliff.

weathering The in-place breakup of any Earth material by physical or chemical means; part of erosion.

Selected Web Sites for
More Geologic Information

Geographic Names
U.S. Geological Survey: www.geonames.usgs.gov/ (Information on about 2 million physical and cultural geographic features.)

Geologic Information
U.S. Geological Survey: www.geology.usgs.gov/
U.S. Geological Survey, Ask a Geologist:
www.walrus.wr.usgs.gov/ask-a-geologist

Geologic Time Scale
Geological Society of America:
www.geosociety.org/science/timescale/timescl.htm (1999)
U.S. Geological Survey:
www.geology.er.usgs.gov/paleo/geotime/shtml (2001)

Geology and Topography Combined (Contiguous United States)
U.S. Geological Survey, A Tapestry of Time and Terrain:
www.tapestry.usgs.gov/ (Can learn more about the geology of specific places such as the Black Hills and Basin and Range; can select state and physiographic boundaries.)

Hazards
U.S. Geological Survey, General Distribution of Hazards in the Contiguous United States: www.usgs.gov/themes/hazards.html (Earthquakes, volcanoes, landslides, major flooding, hurricanes, and tornadoes.)
PBS Savage Earth Online: www.thirteen.org/savageearth (Tsunami, earthquakes, volcanoes.)

Maps
National Atlas: www.nationalatlas.gov (Can generate desired maps, including general geologic maps.)

Museums
Denver Museum of Nature and Science: www.dmns.org
Field Museum of Natural History, Chicago: www.fmnh.org
Museum of Paleontology, University of California, Berkeley:

www.ucmp.berkeley.edu
National Museum, Washington, D.C.: www.si.edu/nmnhweb.html
Royal Tyrrell Museum, Drumheller, Alberta, Canada:
www.tyrrellmuseum.com/

Photographic Tours
U.S. Geological Survey: www.usgs.gov/picturingscience

Photographs, Aquisitioning, General
American Geological Institute, Earth Science World Imagebank:
www.earthscienceworld.org/imagebank
U.S. Geological Survey, Earth Science Photographic Archive:
www.libraryphoto.er.usgs.gov/ (Photographs in the public domain;
can download.)

Photographs, Aquisitioning, Aerial
National Air Photo Library, Ottawa, Canada: NAPL@NRCan.gc.ca
U.S. Geological Survey:
www.edc.usgs.gov/products/aerial.napp.html

Plate Tectonics
U.S. Geological Survey:
www.pubs.usgs.gov/publications/text/dynamic.html

Virtual Field Trips
Hawaii: www.satlab.hawaii.edu/space/hawaii/virtual.field.trips.html
(Aerial and ground photos on six islands.)

Parting Geo-shots

I leave you with four images of Earth features to ponder. Perhaps these, and others in this book, will lead you to further explore our natural world that continues to intrigue geologists. I hope Bare Bones Geology might encourage you to pick at a few more bones.

Parting Geo-shot 1. Grooves in sandstone. Southwest of Rhame, southwestern North Dakota. These grooves, or flutes as they are sometimes called, were cut by overland flow of rainwater into this easily erodible sandstone. Such grooves, with intervening ridges, form interesting badlands slopes in dry terrain. The field of view is about 3 feet (0.9 meter) wide.

Parting Geo-shot 2. Mammoth tooth. Near Millarton, southeastern North Dakota. The tooth, with its grinding surface shown here, is about 6 inches (15 centimeters) high. Elephant-like mammoths became extinct less than 10,000 years ago.

Parting Geo-shot 3. Sunflower-like fossil. Age: Ordovician. Garson, southeastern Manitoba, Canada. This flattened, bowl-shaped fossil consists of squarish or rhomb-like plates in double, intersecting spirals. Inside, the plates attach to rods. Technically *Receptaculites* (ree-sep-tack-you-LIE-teez), this fossil is considered an alga but has been called a sponge and a coral. The fossil, as shown here, is about 6 inches (15 centimeters) wide.

Parting Geo-shot 4. Rock pillar. Spider Rock, Canyon de Chelly, east of Chinle, northeastern Arizona. The pillar, of Permian-age sandstone, was once connected to the rock wall in the upper right of the photograph.

Index

www.ingramcontent.com/pod-product-compliance
Lightning Source LLC
Chambersburg PA
CBHW032019170526
45157CB00002B/765